Statistical Methods for
Comparative Studies

Statistical Methods for Comparative Studies

Techniques for Bias Reduction

SHARON ANDERSON
ARIANE AUQUIER
WALTER W. HAUCK
DAVID OAKES
WALTER VANDAELE
HERBERT I. WEISBERG
with contributions from
ANTHONY S. BRYK
JOEL KLEINMAN

JOHN WILEY & SONS

New York • Chichester • Brisbane • Toronto

Library of Congress Cataloging in Publication Data

Main entry under title:
 Statistical methods for comparative studies.

 (Wiley series in probability and mathematical statistics)
 Includes bibliographical references and index.
 1. Mathematical statistics. I.   Anderson, Sharon, 1948–
 II. Title: Bias reduction.

QA276.S783   001.4'22   79-27220
ISBN 0-471-04838-0

Printed in the United States of America

10 9 8 7 6 5 4 3 2 1

Preface

Investigators in many fields need methods for evaluating the effectiveness of new programs or practices involving human populations. To determine whether a program is more effective than the status quo or another alternative, we must perform comparative studies. An ideal study would apply the different programs to identical groups of subjects. Randomized experiments are often advocated as approximating this ideal. Often, however, randomization is not feasible, resulting in difficult problems of design and analysis. To address these problems, a variety of statistical methods have been developed. Many of these methods are quite recent, and to date have appeared only in technical journals. Although they are potentially very useful to researchers in many fields, these techniques are presently not readily accessible.

In this book we bring together for the first time the various techniques for the design and analysis of comparative studies. The book includes, at a relatively nontechnical level, both familiar techniques and more recent developments. Although we present theoretical results concerning the performance of the various techniques, we emphasize primarily practical implications for the applied researcher. Throughout the book we develop for the applied research worker a basic understanding of the problems and techniques and avoid highly mathematical presentations in the main body of the text.

Overview of the Book

The first five chapters discuss the main conceptual issues in the design and analysis of comparative studies. We carefully motivate the need for standards of comparison and show how biases can distort estimates of treatment effects. The relative advantages of randomized and nonrandomized studies are also presented.

Chapters 6 to 10 present the various methods: matching (including multivariate matching); standardization and stratification; analysis of covariance; and two relatively new multivariate methods, logit analysis and log-linear analysis. We emphasize the assumptions under which the techniques were developed and, whenever possible, assess quantitatively their effectiveness in reducing bias. Although we emphasize estimation as opposed to hypothesis testing, we do indicate the appropriate tests and provide references.

Chapter 11, on survival analysis, deals with the special problem of subject losses during the course of a study and discusses how to form estimates which are not biased by these losses. Chapter 12 discusses repeated measures designs, where subjects are assessed on the same variable both before and after treatment intervention, and presents new methods to handle these problems.

An important feature of the book is Chapter 13. In this chapter we describe the comparative effectiveness of the techniques in reducing bias. In addition, we discuss methods that combine features of two or more techniques. Chapter 14 deals with many of the practical issues that must be faced before drawing causal inferences from comparative studies.

Use of the Book

The book is intended for students, researchers, and administrators who have had a course in statistics or the equivalent experience. We assume that the reader has a basic familiarity with such techniques as regression and analysis of variance, in addition to the basic principles of estimation and hypothesis testing. Depending on the reader's background, some of the relatively more technical sections may be too difficult. The book is written, however, so that the more technical sections can be skipped without loss of understanding of the essentials.

We view this book as serving two different functions. First, the book can be used in a course in research and evaluation methods for students in fields such as public health, education, social welfare, public safety, psychology, medicine, and business. Second, the book serves as a reference for applied researchers wishing to determine which techniques are appropriate for their particular type of study.

However this book is used, we encourage the reader to begin with the first five chapters, because these chapters provide the definitions and lay the foundation for a clear understanding of the problems. A knowledge of the terminology is particularly important, because the fields of application and the statistical literature tend to lack a common terminology. The reader could then refer to Chapter 13, which serves to identify the most appropriate technique(s) (see especially Table 13.1).

Acknowledgments

Work on this book began in the fall of 1974 with discussions in the Observational Studies Group of the Faculty Seminar on the Analysis of Health and Medical Practices at Harvard University. This seminar was supported in part by grants from the Robert Wood Johnsons Foundation; the Edna McConnell Clark Foundation; and the Commonwealth Fund through the Center for the Analysis of Health Practices, Harvard School of Public Health. We wish to thank the organizers and sponsors of this seminar for providing us with a stimulating environment in which to begin our project.

We are indebted to Richard Light for suggesting the idea for this book, starting us in the right direction, and contributing early drafts. Joel Kleinman deserves thanks for early drafts of the first five chapters and helpful comments. We also wish to acknowledge the contribution of Anthony Bryk, who assumed major responsibility for Chapter 12.

Among colleagues who have read drafts and helped with comments, special thanks go to William Cochran, Ted Colton, Allan Donner, William Haenszel, and John Pratt. We are especially grateful to Frederick Mosteller, who provided us with very helpful comments on several drafts and facilitated the production of this book through National Science Foundation Grant SOC-75-15702.

The preparation of this work was partly facilitated by National Science Foundation Grants SOC-75-15702 and SOC-76-15546 (W.V.), and National Institute of Education Contract C-74-0125 (H.W.) and Grant G-76-0090 (H.W.).

SHARON ANDERSON
ARIANE AUQUIER
WALTER W. HAUCK
DAVID OAKES
WALTER VANDAELE
HERBERT I. WEISBERG

Chicago, Illinois
Paris, France
Chicago, Illinois
London, England
Cambridge, Massachusetts
Boston, Massachusetts
July, 1980

Contents

CHAPTER 1

Introduction

This book is concerned with the design and analysis of research studies assessing the effect on human beings of a particular *treatment*. We shall assume that the researchers know what kinds of effects they are looking for and, more precisely, that there is a definite *outcome* of interest. Examples of such treatments and the corresponding outcomes include the administration of a drug (treatment) claimed to reduce blood pressure (outcome), the use of seat belts (treatment) to reduce fatalities (outcome) among those involved in automobile accidents, and a program (treatment) to improve the reading level (outcome) of first graders. As is seen from these examples, the word "treatment" is used in a very general sense.

1.1. PROBLEMS OF COMPARATIVE STUDIES: AN OVERVIEW

It is useful to begin with what might at first sight appear to be an obvious question: What do we mean by the effect of a treatment? We would like to ascertain the differences between the results of two studies. In the first study we determine what happens when the treatment is applied to some group, in the second we determine what would have happened to the same group if it had not been given the treatment of interest. Whatever differences there may be between the outcomes measured by the two studies would then be direct consequences of the treatment and would thus be measures of its effect.

This ideal experiment is, of course, impossible. Instead of doing the second study, we establish a standard of comparison to assess the effect of the treatment. To be effective, this standard of comparison should be an adequate proxy for the performance of those receiving the treatment—the *treatment group*—if they had not received the treatment. One of the objectives of this book is to discuss how to establish such standards of comparison to estimate the effect of a treatment.

Standards of comparison usually involve a *control* or *comparison group* of people who do not receive the treatment. For example, to measure the effect of wearing seat belts on the chance of surviving an automobile accident, we could look at drivers involved in auto accidents and compare the accident mortality of those who wore seat belts at the time of the accident with the accident mortality of those who did not. Drivers who were wearing seat belts at the time of the accident would constitute the treatment group, those who were not would constitute the control group. Ideally, the accident mortality of the control group is close to what the accident mortality of the treatment group would have been had they not worn seat belts. If so, we could use the accident mortality of the control group as a standard of comparison for the accident mortality of the treatment group.

Unfortunately, the use of a control group does not in itself ensure an adequate standard of comparison, since the groups may differ in factors other than the treatment, factors that may also affect outcomes. These factors may introduce a bias into the estimation of the treatment effect. To see how this can happen, consider the seat belt example in more detail.

Example 1.1   Effect of seat belts on auto accident fatality: Consider a hypothetical study attempting to determine whether drivers involved in auto accidents are less likely to be killed if they wear seat belts. Accident records for a particular stretch of highway are examined, and the fatality rate for drivers wearing seat belts compared with that for drivers not wearing seat belts. Suppose that the numbers of accidents in each category was as given in Table 1.1.

From Table 1.1, the fatality rate among drivers who wore seat belts was $10/50 = 0.2$

Table 1.1   Hypothetical Auto Accident Data

	Seat Belts		
	Worn	Not Worn	Total
Driver killed	10	20	30
Driver not killed	40	30	70
Total	50	50	100
Fatality rate	0.2	0.4	

Table 1.2   Auto Accident Data Classified by Speed at Impact

	Low Impact Speed			High Impact Speed		
	Seat Belts Worn	Seat Belts Not Worn	Total	Seat Belts Worn	Seat Belts Not Worn	Total
Driver killed	4	2	6	6	18	24
Driver not killed	36	18	54	4	12	16
Total	40	20	60	10	30	40
Fatality rate	0.1	0.1		0.6	0.6	

and the rate among those not wearing seat belts was $20/50 = 0.4$. The difference of 0.4 $- 0.2 = 0.2$ between the two rates can be shown by the usual chi-square test to be statistically significant at the .05 level. At first sight the study appears to demonstrate that seat belts help to reduce auto accident fatalities.

A major problem with this study, however, is that it takes no account of differences in severity among auto accidents, as measured, for example, by the speed of the vehicle at impact. Suppose that the fatalities among accidents at low speed and at high speed were as given in Table 1.2.

Notice that adding across the cells of Table 1.2 gives Table 1.1. Thus $10 = 6 + 4$, $20 = 2 + 18$, $40 = 36 + 4$, and $30 = 18 + 12$. However, Table 1.2 tells a very different story from Table 1.1. At low impact speed, the fatality rate for drivers wearing seat belts is the same as that for drivers not wearing seat belts, namely 0.1. The fatality rate at high impact speed is much greater, namely 0.6, but is still the same for belted and unbelted drivers. These fatality rates suggest that seat belts have no effect in reducing auto accident fatalities.

The data of Example 1.1 are hypothetical. The point of the example is not to impugn the utility of seat belts (or of well-conducted studies of the utility of seat belts) but to illustrate how consideration of an extra variable (speed at impact) can completely change the conclusions drawn.

A skeptical reader might ask if there is a plausible explanation for the data of Table 1.2 (other than that the authors invented it). The crux of the example is that drivers involved in accidents at low speed are more likely to be wearing seat belts than those involved in accidents at high speed. The proportions, calculated from the third line of Table 1.2, are 40/60 and 10/40, respectively. Perhaps slow drivers are generally more cautious than are fast drivers, and so are also more likely to wear seat belts.

We say that speed at impact is a *confounding factor* because it confounds or obscures the effect, if any, of the risk factor (seat belts, or the lack of them) on outcome (death or survival). In other words, the confounding factor results in a *biased* estimate of the effect.

Fortunately, if (as in Example 1.1) the confounding factor or factors can be identified and measured, the bias they cause may be substantially reduced or even eliminated. Our purpose in this book is to present enough detail on the various statistical techniques that have been developed to achieve this bias reduction to allow researchers to understand when each technique is appropriate and how it may be applied.

1.2 PLAN OF THE BOOK

In Chapters 2 and 3 we discuss the concepts of bias and confounding. In Chapter 3 we also consider the choice of the summary measure used to describe the effect of the treatment. In Example 1.1 we used the difference between the fatality rates of the belted and unbelted drivers to summarize the apparent effect of the treatment, but other choices of measure are possible, for example the ratio of these rates.

The construction of standards of comparison is the subject of Chapter 4. As we have said, these usually involve a control or comparison group that does not receive the treatment. When the investigator can choose which subjects enter the treatment group and which enter the control group, randomized assignment of subjects to the two groups is the preferred method. Since randomization is often not feasible in studies of human populations, we discuss both randomized and nonrandomized studies. In nonrandomized studies statistical techniques are needed to derive valid standards of comparison from the control group, which, as we have seen in Example 1.1, may otherwise give misleading results. Although randomized studies are less likely to mislead, their precision can often be improved by the same statistical techniques.

Chapter 5 discusses the choice of variables to be used in the analysis, a choice that must be related to the context and aims of the study. We also show how the specification of a mathematical model relating the chosen variables is crucial to the choice of an appropriate method of analysis and consider the effects of inadequacies in the model specification.

Chapters 6 to 10 each consider one statistical technique for controlling bias due to confounding factors. These techniques fall into two major categories, *matching* and *adjustment*.

In matching (Chapter 6), the members of the comparison group are selected to resemble members of the treatment group as closely as possible. Matching can be used either to assemble similar treatment and control groups in the planning of the study before the outcomes are determined, or to select comparable subjects from the two groups after a treatment has been given and outcomes measured. Unlike randomization, which requires control over the composition of both groups, matching can be used to construct a comparison group similar to a preselected or self-selected treatment group.

The other major category, adjustment techniques, consists of methods of analysis which attempt to estimate what would have happened if the treatment and comparison groups had been comparable when in fact they were not. In other words, the estimate of the effect of the treatment is adjusted to compensate for the differences between the groups. These adjustment methods include standardization and stratification (Chapter 7), analysis of covariance (Chapter 8), logit analysis (Chapter 9), and log-linear analysis (Chapter 10).

A common problem with longitudinal studies is that subjects may be lost to follow-up at the end of or during the course of the study. Chapter 11, on survival analysis, discusses the analysis of such studies, including the control of confounding factors. Chapter 12 discusses repeated measures designs, where the same subjects are assessed on the outcome variable before and after the intervention of a treatment.

Two summary chapters conclude the book. Chapter 13 discusses the choice of statistical technique and shows how two techniques can sometimes be used together. Finally, Chapter, 14 presents criteria to consider in drawing causal inferences from a comparative study.

The methodological Chapters (6 to 12) may be read in any order, but they all use material from Chapters 1 to 5. Chapter 13 refers in detail to Chapters 6 to 10. Chapter 14 may be read at any point.

The book presents the general rationale for each method, including the circumstances when its use is appropriate. The focus throughout is on unbiased, or nearly unbiased estimation of the effect of the treatment. Tests of significance are given when these can be performed easily. Although we give many examples to illustrate the techniques, we do not dwell on computational details, especially when these can best be performed by computer. We shall assume throughout the book that the researchers have chosen a single outcome factor for study. For simplicity of presentation we often also restrict attention to the estimation of the effect of a single treatment in the presence of a single confounding factor, although extensions to multiple confounding factors are indicated. Some special issues that arise with multiple confounding factors are discussed in Chapter 5.

Throughout the book the main concern will be *internal validity*—attaining a true description of the effect of the treatment on the individuals in the study. The question of *external validity*—whether the findings apply also to a wider group or population—is not discussed in depth as it is primarily determined by the subject matter rather than by statistical considerations.

1.3   NOTES ON TERMINOLOGY

Throughout, we shall refer to the effect of interest as the *outcome factor.* A common synonym is *response factor.* The agent whose effect on the outcome factor is being studied will be called the *treatment, treatment factor,* or *risk factor.* The word "treatment" is generally used to describe an agent applied specifically to affect the outcome factor under consideration (as was true for all the examples in the first paragraph of this chapter). The term "risk factor," borrowed from epidemiology, is used when exposure to the agent is accidental or uncontrollable, or when the agent is applied for some purpose other than to affect the specific outcome factor under consideration. An example would be the study of the effect of smoking on the incidence of lung cancer. The use of the term "risk factor" does not in itself imply that the agent is "risky" or in fact, that risk enters the discussion at all. We use whichever term ("treatment" or "risk factor") appears more natural in context.

In later chapters we talk about quantities or labels that measure the presence, absence, level or amount of a risk factor, treatment, outcome factor, or confounding factor. Such quantities or labels will be termed *variables.* In studying the effect of seat belts on accident mortality (Example 1.1) we may define a risk variable taking the value 1 or 0, depending on whether or not the driver was wearing a seat belt at the time of the accident. The logical distinction between a factor and a variable which measures that factor is not always made in the literature, but it can be useful.

The term "comparison group" is used interchangeably with the more familiar "control group." When the important comparison is between a proposed new treatment and the present standard treatment, the standard treatment (rather than no treatment) should be given to the comparison group. In dealing with risk factors it is natural to speak of "risk groups" or of "exposed" and "nonexposed" groups. We may have several different "exposed" or "treatment" groups, corresponding to different levels of the risk factor or treatment.

CHAPTER 2

Confounding Factors

In the discussion of Example 1.1 (effect of wearing seat belts on auto accident fatality) we saw that a background factor (speed at impact) could seriously distort the estimate of the effect of the risk factor on the outcome. The distortion will arise whenever two conditions hold:

1. The risk groups differ on the background factor.
2. The background factor itself influences the outcome.

Background factors which satisfy conditions 1 and 2 are called confounding factors. If ignored in the design and analysis of a study, they may affect its conclusions, for part of the effect of the confounding factor on the outcome may appear to be due to the risk factor. Table 1.1 is misleading because the effect on accident fatality apparently due to wearing seat belts (the risk factor) is actually due to speed at impact (the confounding factor).

In Section 2.1 we show by another example how the effect of a risk factor can sometimes be disentangled from that of a confounding factor. A useful measure of the likely influence of a confounding factor on the estimate of treatment effect is the *bias*. Section 2.2 quantifies the term "bias" and briefly introduces the concepts of precision and statistical significance. The qualitative discussion in Chapter 1 of the relation among the risk, outcome, and confounding factors is extended in Section 2.3. Formulas relating bias, standard error, and mean squared error are given in Appendix 2A.

2.1 ADJUSTMENT FOR A CONFOUNDING FACTOR

In Example 1.1 the risk factor had no actual effect on the outcome. Table 1.2 shows that its apparent effect was due entirely to the confounding factor. In most studies many factors will each have some effect on the outcome and the investigator will want to estimate the magnitude of the treatment effect after allowing for the effect of the other factors. An example will show that sometimes this can be done quite easily.

Example 2.1 Coffee drinking, obesity, and blood pressure: Suppose that a physician, Dr. A, wants to assess the effect on the diastolic blood pressure of his male patients of their regularly drinking coffee. We shall consider just two levels of the risk factor, coffee drinking, corresponding to patients who drink coffee regularly (the drinkers) and patients who do not drink coffee regularly (the nondrinkers). The outcome variable, diastolic blood pressure, is a numerical measurement. Dr. A is unwilling to instruct his patients to drink coffee or to stop drinking coffee, but he can rely (let us say) on truthful answers to questions on the subject in his medical records.

Because he knows that blood pressure is also influenced by weight—overweight patients tend to have higher blood pressures that those of normal weight—Dr. A classifies all his male patients by obesity (overweight or not overweight) as well as by coffee drinking. Dr. A calculates the average diastolic blood pressure in millimeters of mercury (mm Hg) of patients in the four categories. We shall suppose that the average diastolic blood pressure among the nondrinkers who are not overweight is 70 mm Hg, but that among the nondrinkers who are overweight the average is 90 mm Hg. Let us also suppose that the effect of drinking coffee regularly is to increase blood pressure by exactly 4 mm Hg, and that there are no other complicating factors. Then the average diastolic blood pressures among the drinkers who are and who are not overweight are 94 and 74 mm Hg, respectively. These assumptions are summarized in Table 2.1. Notice that we have not yet specified the numbers of patients in each category.

Suppose that Dr. A were to attempt to estimate the effect of drinking coffee on blood

Table 2.1 *Average Diastolic Blood Pressures (mm Hg)*

	Overweight	Not Overweight
Drinkers	94.0	74.0
Nondrinkers	90.0	70.0

Table 2.2   *"Even" Distribution for Dr. A's Patients*

	Overweight	Not Overweight	Total
Drinkers	100	300	400
Nondrinkers	50	150	200

pressure ignoring the effect of obesity. He would compare the average blood pressure of the drinkers with that of the nondrinkers. To calculate these averages Dr. A will need to know the numbers of his patients in each category of Table 2.1. We shall suppose that he has 600 male patients in all, and will consider two different distributions of their numbers, an "even" distribution (Table 2.2) and an "uneven" distribution (Table 2.3).

In the "even" distribution the proportion of overweight patients among the drinkers (100/400 = 0.25) is the same as that among the nondrinkers (50/200 = 0.25). In statistical language, Table 2.2 exhibits no association between coffee drinking and obesity. The average blood pressure among all the drinkers is the weighted mean of the averages on the top line of Table 2.1, weighted by the numbers of patients contributing to each average. From Table 2.1 and Table 2.2, this is

$$\frac{(94.0 \times 100) + (74.0 \times 300)}{100 + 300} = 79.0 \text{ mm Hg.}$$

From the second line of the same tables, the average blood pressure among the nondrinkers is

$$\frac{(90.0 \times 50) + (70.0 \times 150)}{50 + 100} = 75.0 \text{ mm Hg.}$$

Dr. A's estimate of the average increase in blood pressure due to coffee drinking would be

$$79.0 - 75.0 = 4.0 \text{ mm Hg.}$$

This is the correct answer because it agrees with the rise of 4.0 mm Hg that we assigned to coffee drinking. To summarize, if there is no association between the risk factor, coffee drinking, and the background factor, obesity, among Dr. A's patients, a straight comparison of average blood pressures among the drinkers and among the nondrinkers will be adequate. Here the background factor satisfies condition 2 of the definition of a confounding factor given at the beginning of this chapter, but it does not satisfy condition 1 and so is not a confounding factor.

If, instead, Dr. A's patients follow the "uneven" distribution of Table 2.3, then both parts of the definition will be satisfied, as Table 2.3 does indicate an association between coffee drinking and obesity. Obesity will now be a confounding factor. The average blood

Table 2.3   *"Uneven" Distribution for Dr. A's Patients*

	Overweight	Not Overweight	Total
Drinkers	300	100	400
Nondrinkers	50	150	200

pressure among the drinkers who visit Dr. A will be

$$\frac{(94.0 \times 300) + (74.0 \times 100)}{300 + 100} = 89.0 \text{ mm Hg.}$$

Among the nondrinkers, the average blood pressure will be

$$\frac{(90.0 \times 50) + (70.0 \times 150)}{50 + 150} = 75.0 \text{ mm Hg.}$$

The crude estimate of the average increase in blood pressure due to coffee drinking, namely

$$89.0 - 75.0 = 14.0 \text{ mm Hg,}$$

would be incorrect.

Of course, this problem does not arise if Dr. A assesses the effect of coffee drinking separately among his overweight patients and among his patients who are not overweight. He then uses the values given in Table 2.1 to arrive at the correct estimate of the effect of coffee drinking, namely that it increases average blood pressure by 4.0 mm Hg among both classes of patient.

However, Dr. A may prefer to calculate a single summary measure of the effect of coffee drinking on blood pressure among all his patients. He can do this by applying the average blood pressures in Table 2.1 to a single hypothetical standard population consisting, for example, of 50% patients of normal weight and 50% patients who are overweight. These calculations would tell him what the average blood pressures would be in this standard population (a) if they all drank coffee and (b) if none of them drank coffee. The calculations give

$$(94.0 \times 0.50) + (74.0 \times 0.50) = 84.0 \text{ mm Hg}$$

for the average blood pressure among the patients in the standard population if they were all to drink coffee, and

$$(90.0 \times 0.50) + (70.0 \times 0.50) = 80.0 \text{ mm Hg}$$

if none of them were to drink coffee. The comparison between these two averages gives the correct result.

This adjustment procedure is an example of standardization, to be described further in Chapter 7.

2.2   BIAS, PRECISION, AND STATISTICAL SIGNIFICANCE

For many reasons the estimated treatment effect will differ from the actual treatment effect. We may distinguish two types of error: random error, and systematic error or *bias*. This book is primarily concerned with the bias introduced into the estimate of treatment effect by confounding factors. However, to illustrate the distinction between random error and bias, we give a simple example not involving a treatment or confounding factor.

Example 2.2   The Speak-Your-Weight machines: An old Speak-Your-Weight machine is rather erratic but not discernibly off-center. A new machine gives perfectly reproducible results but may be off-center because of an incorrect setting at the factory.

A man of weight 170 lb who weighs himself five times on the old machine may hear weights of 167, 172, 169, 173, and 168 lb. The new machine might respond 167, 167, 167, 167, and 167 lb. The old machine is exhibiting random error, the new machine systematic error. Of course, a Speak-Your-Weight machine could easily be both erratic and off-center. Such a composite machine, exhibiting the defects described both for the old and the new machines, might give readings of 164, 169, 166, 170, and 165 lb, which are subject to random error and to bias.

There is clearly no way to distinguish between these types of error by a single measurement (e.g., of 167 lb). Implicit in the distinction between random error and systematic error is the notion of repetition: random error would approximately cancel out if repeated measurements were taken and averaged [in this example the average of the five weights spoken by the old machine, $(167 + 172 + 169 + 173 + 168)/5 = 169.8$ lb is quite close to the true weight, 170 lb], while systematic error is impervious to averaging (the average of the weights spoken by the new machine is still 167 lb).

Statistical techniques such as significance testing and the calculation of standard errors and confidence intervals are often helpful in gauging the likely effect of random error on the conclusions of a study. These techniques cannot in themselves assess the effect of systematic error.

2.2.1 Bias

We can now attempt a formal definition of the term "bias."

Definition: The *bias* of an estimator is the difference between the average value of the estimates obtained in many repetitions of the study and the true value of what it is estimating.

By thinking of an estimator as a procedure that produces estimates, we introduce the notion of repetition into the definition. Because of random error the estimate would change from repetition to repetition, although the estimator, the procedure used to derive the estimates, would not change. The definition emphasizes that the bias is a number, positive or negative. This contrasts with the common use of the term as an abstract noun, or even as a general insult to impugn any study that disagrees with one's own opinions: "This study is biased."

Confounding factors are the major source of bias in nonrandomized studies (in both the common and the technical usage of the term "bias") and it is with the bias due to confounding factors that this book is primarily concerned. Other possible sources of bias will be mentioned in Chapter 5.

Unfortunately, the definition we have just given rarely enables the bias to be calculated, even in terms of the unknown true treatment effect, since studies are not repeated and we cannot say what would happen if they were. With the partial exceptions of matching and standardization, all the techniques described herein depend on assumed *statistical models,* which state what would happen in hypothetical repetitions. A simple statistical model for the weight X registered by the "old" Speak-Your-Weight machine of Example 2.2 has

$$X = \mu + \epsilon,$$

where μ is the true weight of the man and ϵ denotes a random error. The true weight μ would not change from repetition to repetition, but the random error ϵ would change, with an average value close to zero after many repetitions.

By contrast, the "new" Speak-Your-Weight machine has

$$X = \mu + b,$$

where μ is the true weight, as before, and b is a systematic error which does not change from repetition to repetition.

Rarely can the validity of an assumed statistical model be checked directly. The methodological chapters (Chapters 6 to 12) will discuss the statistical models demanded by each technique and such indirect checks of the validity of these models as are available. The equations are not usually as simple as those given above because they must relate the outcome variable to the treatment and confounding variables of interest and to the measure chosen to describe the effect of the treatment. The distribution of the random error must also be specified.

2.2.2 Precision and Statistical Significance

The precision of an unbiased estimator of a treatment effect is usually measured by the variance of the estimator or by the square root of this variance, the standard error. The smaller the variance or standard error, the more precise is the estimator. The standard error of a biased estimator still measures the influence of random error on the estimator, but it gives no clue as to the magnitude of systematic error. As systematic error is usually a more serious threat to the validity of observational studies than is random error, this book assesses techniques by their ability to reduce bias and places only a secondary emphasis on precision. However, most of the procedures we describe permit the calculation of standard errors of estimated treatment effects.

The *mean squared error* of an estimator is defined as the mean value, in hypothetical repetitions, of the square of the difference between the estimate and the true value. We show in Appendix 2A that the mean squared error can be calculated as the variance plus the square of the bias. It provides a useful criterion for the performance of estimators subject to both systematic and random error.

The function of a test of *statistical significance* is to determine whether an apparent treatment effect could reasonably be attributed to chance alone. When applied to data from well-designed randomized studies, significance tests can effectively demonstrate the reality of the observed treatment effect. In non-randomized studies, where systematic error will usually provide a more plausible explanation of an observed treatment effect than will random variation, significance tests are less crucial. Nevertheless, they can, if carried out after adjustment for confounding factors, be useful indicators of whether the observed treatment effect is real.

The concepts of precision and statistical significance are closely related. Whether an estimated treatment effect is statistically significant depends not only on the magnitude of the estimated effect but also on the precision of the estimator. A useful rule of thumb, based on an assumed normal distribution, holds an estimated treatment effect at least twice its standard error from the no-effect value to be on the borderline of statistical significance, and to be highly significant if away by at least three times its standard error.

The methodological chapters include some discussion of tests of statistical significance and of the precision of estimators.

2.3 SOME QUALITATIVE CONSIDERATIONS

For the two examples involving confounding factors discussed so far (seat belts to reduce accident fatalities, effect of coffee drinking on blood pressure), the assumed relations among the factors are summarized in Figures 2.1 and 2.2.

In these figures an arrow ($\rightarrow$) denotes a direct casual link. That is, $A \rightarrow B$ if a change in A would result in a change in B if all other factors listed in the figure do not change. A double arrow ($\leftrightarrow$) denotes a possible association between factors A and B which may not have a simple causal interpretation. The two factors may influence each other and may both be influenced by other factors not included in the figure. The relation of primary interest is, as always, that

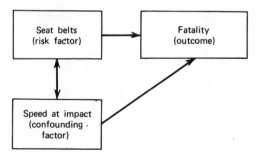

Figure 2.1 *Seat belts and fatalities.*

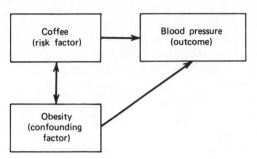

Figure 2.2   *Coffee drinking and blood pressure.*

between the risk factor and the outcome. The figures indicate the defining properties of a confounding factor: it is associated with the risk factor and it influences the outcome. As we have seen, the correct statistical analysis for both Examples 1.1 and 1.2 is to adjust for the effect of the confounding factor.

2.3.1   Unnecessary Adjustment

The following example, from MacMahon and Pugh (1970, p. 256), suggests that adjustment is not always called for.

Example 2.3   Oral contraceptives and thromboembolism: Consider an investigation of the effect of oral contraceptives on the risk of thromboembolism in women. A factor possibly associated with the risk factor (use of oral contraceptives) is religion. Catholic women may be less likely to use oral contraceptives than are other women. The relation between the three factors mentioned might be as shown in Figure 2.3. The cynic may add a second arrowhead to the arrow connecting "Religion" and "Oral contraceptive." As always, the relation between the risk factor (oral contraceptive use) and the outcome (thromboembolism) is of primary interest.

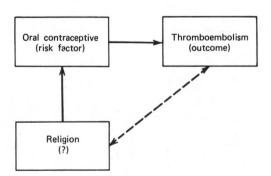

Figure 2.3   *Oral contraceptives and thromboembolisms.*

Table 2.4   Number of Thromboembolisms and Number of Women by Religion and Oral Contraceptive (OC) Use

	Catholic	Non-Catholic	Total
OC users	2000	5000	7000
(thromboembolisms)	(100)	(250)	(350)
Nonusers	8000	5000	13,000
(thromboembolisms)	(240)	(150)	(390)

To amplify the discussion, let us assume that the true lifetime risks of thromboembolism among users and nonusers of the contraceptive pill are 5% and 3%, respectively, irrespective of religion. Consider a study population consisting of 10,000 Catholic women and 10,000 non-Catholic women and suppose that 20% of the Catholics but 50% of the non-Catholics use oral contraceptives. Table 2.4 gives the number of women in each category of the study population and the number of these women who would suffer a thromboembolism if the rates of 5% and 3% were to apply.

In this example an analysis ignoring religion will give the correct risks (350/7000 = 0.05 and 390/13,000 = 0.03), as should be clear from the construction of Table 2.4. However, the background factor of religion is apparently related not only to the risk factor—this we assumed at the start—but also to the outcome, as Table 2.5 demonstrates. The risk of thromboembolism is slightly higher among non-Catholics than among Catholics. Apparently, religion here satisfies the definition of a confounding factor, since it is a background factor associated with both the risk factor and the outcome.

Closer examination reveals that religion does not satisfy the definition. Although this background factor is associated with the outcome, it does not influence the outcome except through its effect on the risk factor. The dashed arrow in Figure 2.3 is a consequence of the other two arrows in the diagram.

If, nevertheless, the investigator does choose to correct for religion as a confounding factor using one of the techniques described in later chapters, he or she will not introduce bias into the study. Depending on the procedure chosen, there will be a slight or substantial loss of precision.

This last point applies more generally. Unnecessary adjustment—adjustment for a background factor that is not in fact confounding—will not introduce bias into a study except in some rather special circumstances, involving regression effects to be discussed in Section 5.3 (but note also Example 2.4). However, the precision of the estimated treatment effect may be reduced.

Table 2.5   Totals from Table 2.4

	Catholic	Non-Catholic
All women	10,000	10,000
Thromboembolisms	(340)	(400)

2.3.2 Proxy Variables

Before leaving Example 2.3, we should consider the possible effects of other important background variables. In fact, correction for the effect of religion will be useful if religion is associated with a confounding variable not measured in the study. Religion would then be called a *proxy variable*. This could happen, for example, in the following cases:

1. If risk of thromboembolism is affected by diet and the eating habits of Catholic and non-Catholic women differ. Diet would then be confounding, being related to both the risk factor (oral contraceptive use), through its relation to religion, and to the outcome (thromboembolism).

2. If risk of thromboembolism is affected by family size, and Catholic women had more children than did non-Catholic women. Here family size would be confounding for the same reason as diet in (1).

The investigator may choose to adjust for religion as a substitute for the un-measured confounding factor. Unfortunately, the association between the proxy variable and the unmeasured confounding factor needs to be quite strong before the former can substitute effectively for the latter.

2.3.3 Defining the Factors

In some situations confusion over the definition of the risk factor can actually introduce bias into the study.

Example 2.4 Maternal age and infant mortality: Suppose that we want to determine the effect of maternal age on infant mortality. Birth weight might be considered as a confounding factor, as older mothers have lower-weight babies and lower-weight babies have higher mortality. However, adjusting for birth weight in the analysis would be

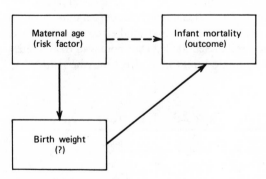

Figure 2.4 *Maternal age and infant mortality.*

misleading, because we would be adjusting away the major difference we should be looking for. Birth weight in this example is a kind of intermediate outcome which leads to the final outcome of interest. Figure 2.4 summarizes the relationships among the three factors. If the effect of maternal age on infant mortality is entirely attributable to its effect on birth weight, an analysis adjusted for birth weight will indicate no association between maternal age and infant mortality.

Of course, it is possible that maternal age affects infant mortality through factors other than birth weight. Two infants of identical birth weight but whose mothers were of different ages would then be subject to different risks. An investigator interested in the effect of these other factors should adjust for birth weight. The new, adjusted estimate of the effect of the risk factor would differ from the unadjusted estimate, because the investigator's definition of the risk factor would be different.

Often the question of whether to adjust for a particular factor is not statistical but arises because the researcher has not defined with sufficient care the risk factor he or she wants to study.

APPENDIX 2A   BIAS, PRECISION, AND MEAN SQUARED ERROR

Let θ denote the true value of the treatment effect and $\hat{\theta}$ the estimator of θ. The expectation symbol E denotes averaging with respect to the distribution of $\hat{\theta}$ in hypothetical repetitions. The bias, variance, and mean squared error (m.s.e.) of $\hat{\theta}$ are, respectively,

$$\text{bias}\ (\hat{\theta}) = E(\hat{\theta}) - \theta$$
$$\text{var}\ (\hat{\theta}) = E[\hat{\theta} - E(\hat{\theta})]^2 = E(\hat{\theta})^2 - [E(\hat{\theta})]^2$$
$$\text{m.s.e.}\ (\hat{\theta}) = E(\hat{\theta} - \theta)^2.$$

On expanding the squared term in the last formula, we see that the cross-product term vanishes, and we obtain

$$\begin{aligned}\text{m.s.e.}\ (\hat{\theta}) &= E[\hat{\theta} - E(\hat{\theta}) + E(\hat{\theta}) - \theta]^2 \\ &= E[\hat{\theta} - E(\hat{\theta})]^2 + [E(\hat{\theta}) - \theta]^2 \\ &= \text{var}\ (\hat{\theta}) + [\text{bias}\ (\hat{\theta})]^2.\end{aligned}$$

REFERENCE

MacMahon, B. and Pugh, T. F. (1970), *Epidemiology Principles and Methods,* Boston: Little, Brown.

CHAPTER 3

Expressing the
Treatment Effect

In an ideal hypothetical situation we could observe on the same group of individuals the outcome resulting both from applying and from not applying the treatment. We could then calculate the effect of the treatment by comparing the outcomes under the two conditions. We could define a measure of treatment effect for each individual as the difference between his or her outcomes with and without the treatment. If all subjects were exactly alike, this measure would be the same for each. But more commonly, differences between subjects will cause the measure to vary, possibly in relation to background factors. A treatment may, for instance, be more beneficial to younger than to older people; so the effect would vary with age. We may then wish to define a summary measure of the effect of the treatment on the entire group.

In Section 3.1 we will explore different summary measures of treatment effect. In Example 2.1, Dr. A's choice was to express the treatment effect as the average difference in blood pressure between patients who drink coffee and those who do not drink coffee. We will see that this choice was dictated partly by the nature of the risk factor and partly by the underlying model that Dr. A had in mind as to how coffee consumption affects blood pressure. In Section 3.2 we will leave our ideal situation and see how, when we use a comparison group to estimate a summary measure of treatment effect, a confounding factor may distort that estimate. Finally, in Section 3.3 we will focus on situations in which the treatment effect is not constant and show that a single summary measure of treatment effect might not be desirable.

3.1   MEASURES OF TREATMENT EFFECT

The choice of measure for treatment effect depends upon the form of the risk and outcome variables. It is useful to make the distinction between a *numerical* variable and a *categorical* variable. The levels of a numerical variable are numbers, whereas the levels of a categorical variable are labels. Thus age expressed in years is a numerical variable, whereas age expressed as young-middle-aged/old or religion expressed as Catholic/Protestant/Jewish/other are categorical variables. Since the levels of a numerical variable are numbers, they can be combined to compute, for instance, a mean (e.g., the mean age of a group of individuals). For categorical variables, on the other hand, the levels are looked at separately (e.g., there are 45 young individuals, 30 middle-aged, and 60 old). Categorical variables with only two possible levels (e.g., intensive reading program vs. standard reading program) are called *dichotomous* variables.

Furthermore, we will sometimes distinguish between an *ordered categorical* variable, such as age, and an *unordered categorical* variable, such as religion. There exists for the first type an intrinsic ordering of the levels (e.g., young/middle-aged/old), whereas for the second type there is no relationship between the levels (e.g., we cannot arrange the various religions in any particular order). A numerical variable can be created from an ordered categorical variable by assigning numbers or *scores* to the different levels (e.g., −1 to young, 0 to middle-aged, and 1 to old).

Using numerical and categorical variables, we can distinguish four different situations, as shown in Figure 3.1. In this book we are concerned mainly with Cases 1 and 2, where the risk variable is categorical.

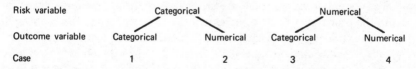

Figure 3.1   *Different cases for measures of treatment effect.*

Case 1:   Consider first the effect of a treatment on a dichotomous outcome, specifically death or survival. Three measures of treatment effect are commonly used (Fleiss, 1973; see also Sheps, 1959, for other proposals). We define the three measures and illustrate their use with the data given in Table 3.1. Notice that

Table 3.1   *Measures of Treatment Effect for Dichotomous Treatment and Outcome in Three Examples*

| | Example (a) | | Example (b) | | Example (c) | |
	Treatment	Control	Treatment	Control	Treatment	Control
Death rate	0.06	0.01	0.55	0.50	0.60	0.10
Survival rate	0.94	0.99	0.45	0.50	0.40	0.90
Difference of death rates (Δ)	$0.06 - 0.01 = 0.05$		$0.55 - 0.50 = 0.05$		$0.60 - 0.10 = 0.50$	
Relative risk (θ)	$0.06/0.01 = 6.00$		$0.55/0.50 = 1.10$		$0.60/0.10 = 6.00$	
Odds ratio (ψ)	$\dfrac{0.06}{0.94} \Big/ \dfrac{0.01}{0.99} = 6.32$		$\dfrac{0.55}{0.45} \Big/ \dfrac{0.50}{0.50} = 1.22$		$\dfrac{0.60}{0.40} \Big/ \dfrac{0.10}{0.90} = 13.50$	

in all three examples given in Table 3.1 the treatment is harmful, since the death rate is higher in the treatment group than in the control group. The three measures of treatment effect are:

- The difference in death rates (Δ) between the treatment and control groups. (In epidemiology this is called the attributable risk.) In example (a) in Table 3.1, $\Delta = 0.05$ means that the risk of dying is 0.05 greater in the treatment group.
- The relative risk (θ) is defined as the ratio of the death rate in the treatment group to the death rate in the control group. In example (c) in Table 3.1, $\theta = 6$ implies that the risk of dying in the treatment group (0.60) is 6 times higher than the risk of dying in the control group (0.10).
- The odds ratio (ψ) or *cross-product ratio* is based on the notion of odds. The odds of an event are defined as the ratio of the probability of the event to the probability of its complement. For instance, the odds of dying in the treatment group of example (c) are equal to the death rate (0.60) divided by the survival rate (0.40), or 1.50. When the odds of dying are greater than 1, the risk or probability of dying is greater than that of surviving. Now, the odds ratio in our example is the ratio of the odds of dying in the treatment group (1.50) to the odds of dying in the control group (0.10/0.90 = 0.11), or 13.50. The odds of dying are 13.50 times higher in the treatment group. The odds ratio can be conveniently computed as the ratio of the product of the diagonal cells of the treatment by survival table—hence its alternative name, cross-product ratio. In our

example,

$$\psi = \frac{0.60}{0.40} \Bigg/ \frac{0.10}{0.90} \qquad \text{(ratio of the odds)}$$

$$\text{or} = \frac{(0.60) \times (0.90)}{(0.40) \times (0.10)} \qquad \text{(cross-product ratio)}$$

$$= 13.50.$$

The three measures of treatment effect—difference of rates (Δ), relative risk (θ), and odds ratio (ψ)—are linked in the following ways:

1. If the treatment has no effect (i.e., the death rates are equal in the control and treatment groups), then $\Delta = 0$ and $\theta = \psi = 1$.

2. If Δ is negative or θ or ψ smaller than 1, the treatment is beneficial. Conversely, if Δ is positive, θ or ψ greater than 1, the treatment is harmful.

3. If the death rates in the treatment and control groups are low, the odds ratio and relative risk are approximately equal [see, e.g., Table 3.1, example (a); see also Appendix 4A].

4. In certain types of studies (see case-control studies in Chapter 4), only the odds ratio can be meaningfully computed. In these studies the total number of deaths and the total number of survivors are fixed by the investigator, so that death rates and hence differences of death rates and relative risks cannot be interpreted. We shall see in Chapter 4 that the odds ratio does have a sensible interpretation in these studies.

The three examples of Table 3.1 were chosen in such a way that (a) and (b) lead to the same difference of rates and (a) and (c) to the same relative risk. These examples show that the value of one of the three measures has no predictable relation (other than those mentioned above) to the value of any other two: although (a) and (b) have the same Δ of 0.05, their relative risks (6.00 and 1.10) are widely different.

Several factors influence the choice of the measure of treatment effect. The choice may depend on how the measure is going to be used. For example, a difference in death rates would give a better idea of the impact that the treatment would have if it were applied to all diseased people (MacMahon and Pugh, 1970). Berkson (1958; also quoted in Fleiss, 1973), in looking at the effect of smoking on survival, makes this point by saying that "of course, from a strictly practical viewpoint, it is only the total number of increased deaths that matters." On the other hand, the relative risk may highlight a relationship between a risk and an outcome factor. Hill (1965) remarks that although 71 per 10,000 and 5 per 10,000 are both very low death rates, what "stands out vividly" is that the first is 14 times the second. Thus the choice of a measure may be guided by the aim of the study.

Also, the investigator may believe that one model is more appropriate than another in expressing how the treatment affects the outcome, and he or she can use the data at hand to test his or her belief. That particular model may suggest a measure of treatment effect. This applies for any of the four cases considered in this section. We will turn to Case 2 and illustrate there how a measure may derive from a model.

Case 2: When the outcome variable is numerical (e.g., weight, blood pressure, test score), the difference of the average of the outcome variable between the treatment and comparison groups is a natural measure of treatment effect. For instance, Dr. A. can calculate the average blood pressure among coffee drinkers and among non-coffee drinkers and take the difference as a measure of treatment effect.

Dr. A. may think of two different ways in which coffee might affect blood pressure. Let Y_1 and Y_0 be the blood pressure of a given patient with and without coffee drinking. First, coffee drinking might increase blood pressure by a certain amount Δ, which is the same for all patients:

$$Y_1 = Y_0 + \Delta \qquad \text{for any patient (ignoring random variation).}$$

Second, coffee drinking might increase blood pressure proportionally to each patient's blood pressure. Let π be this coefficient of proportionality:

$$Y_1 = \pi Y_0 \qquad \text{for any patient.}$$

By taking logarithms on each side of this expression, we have, equivalently,

$$\log Y_1 = \log Y_0 + \log \pi.$$

Notice that we have transformed a multiplicative effect (π) into an additive effect ($\log \pi$) by changing the scale of the variables through the logarithmic function.

In the first case, Δ would be the measure of treatment effect suggested by the model, which Dr. A. could estimate by the difference of average blood pressure in the coffee and no-coffee group. In the second case, he could consider $\log \pi$ as a measure of treatment effect, which he could estimate by the difference of the average logarithm of blood pressure between the two groups. Or he may find π easier to interpret as a measure of treatment effect and transform back to the original units through the exponential function. Clearly, with the data at hand (see Table 2.1), the first model (and hence Δ) is more appropriate.

Case 3: An example of Case 3, where the risk variable is numerical and the outcome categorical, is a study of increasing doses of a drug on the chance of surviving for 1 year. The odds of dying can be defined for each dose of the drug. The effect of the drug can be assessed by looking at the change in the odds of dying as the dose increases. A model often used in such cases assumes that for any increase of the dose by 1 unit, the logarithm of the odds changes by a constant amount. This amount is taken as the measure of treatment effect.

Case 4:   Here both the risk and outcome variables are numerical. Suppose that we want to look at the effect of increasing doses of a drug on blood pressure; if a straight line is fitted to the blood pressure–dose points, the slope of the line can be taken as a measure of the effect of the drug. It represents the change in blood pressure per unit increase in dosage. Regression techniques that can be used in this case will not be discussed in this book. This topic has been covered in many other books (see, e.g., Tufte, 1974; Mosteller and Tukey, 1977; Hanushek and Jackson, 1977; Daniel and Wood, 1971; Colton, 1974).

From the discussion of these four cases, it should be clear that a measure of treatment effect not only depends on the form of the risk and outcome variables, but also on the aim of the study, the scale of the variables, and the models judged appropriate by the investigators.

3.2   WHAT HAPPENS WHEN THERE IS CONFOUNDING

We know from previous chapters that we might be wary of confounding factors when we compare a group of treated individuals and a group of comparison individuals to assess the effect of a treatment. The purpose of this section is to show how a confounding factor distorts the estimate of the treatment effect, and how crude odds ratios or differences of average outcome are not good estimates of treatment effect in the presence of confounding.

As before, we will consider different cases, depending on how the outcome and confounding factors are measured (i.e., whether they are numerical or categorical). We will consider here only dichotomous risk variables, one level being the treatment and the other the comparison. Figure 3.2 illustrates the four possibilities. The numbers (2) and (1) at the top of the figure refer to the case

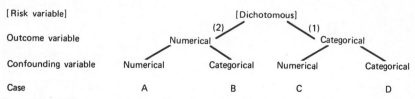

Figure 3.2   *Different cases for the effect of a confounding factor.*

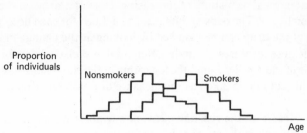

Figure 3.3 *Age distribution in the smoking and nonsmoking groups.*

numbers in Figure 3.1, and these indicate which measures of treatment effect are appropriate for Cases A, B, C, and D.

An example of *Case A* is a study of the effect of smoking on blood pressure where age expressed in years would be a confounding factor. Suppose that the smoking and nonsmoking groups that we compare have the age distributions shown in Figure 3.3. Note that there are very few young smokers and very few old nonsmokers. The average age of smokers is greater than the average age of nonsmokers.

In addition, suppose that a plot of blood pressure vs. age in each group suggests, as in Figure 3.4, that blood pressure is linearly related to age, with equal slopes among smokers and nonsmokers. If we denote blood pressure by Y and age by X and use the subscripts S for smokers and NS for nonsmokers, we have (ignoring random variation)

$$Y_S = \alpha_S + \beta X_S \qquad \text{in the smoking group}$$
$$Y_{NS} = \alpha_{NS} + \beta X_{NS} \qquad \text{in the nonsmoking group.}$$

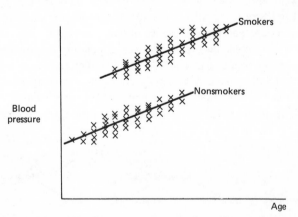

Figure 3.4 *Relationship of blood pressure with age in the smoking and nonsmoking groups.*

The same slope (β) appears in the two equations, but the intercepts α_S and α_{NS} are different.

Note that age satisfies the definition of a confounding factor given in Chapter 2: it has a different distribution in the smoking and nonsmoking groups (Figure 3.3) and it affects blood pressure within each population (Figure 3.4). If we assume that age and smoking are the only factors affecting blood pressure, we can measure the effect of smoking by the vertical distance between the two lines of Figure 3.4 (i.e., $\alpha_S - \alpha_{NS}$).

In the discussion of Case 2 in Section 3.1, we suggested measuring the treatment effect by the difference between the average outcomes: in our example by $\overline{Y}_S - \overline{Y}_{NS}$, the difference between the average blood pressure in the smoking group and that in the nonsmoking group. Since

$$\overline{Y}_S = \alpha_S + \beta\overline{X}_S$$
$$\overline{Y}_{NS} = \alpha_{NS} + \beta\overline{X}_{NS}$$

(where the overbar indicates that we have averaged over the group), it follows that

$$\overline{Y}_S - \overline{Y}_{NS} = (\alpha_S + \beta\overline{X}_S) - (\alpha_{NS} + \beta\overline{X}_{NS})$$
$$= (\alpha_S - \alpha_{NS}) \qquad + \beta(\overline{X}_S - \overline{X}_{NS})$$
$$= \text{treatment effect} + \text{bias}.$$

Thus if we use the difference of average blood pressure, in our example we overestimate the treatment effect by the amount $\beta(X_S - \overline{X}_{NS})$, which we call the bias. We have represented this situation in Figure 3.5, which combines Figures 3.3 and 3.4. (In Figure 3.5 the age distributions in each group from Fig.

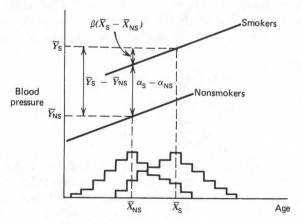

Figure 3.5 *Treatment effect and bias.*

3.3 appear at the bottom of the figure and the relationships between blood pressure and age from Figure 3.4 appear as solid lines. The vertical axis of Figure 3.3 is not explicitly shown.) Note that if age were not a confounding factor, either the age distribution would be the same in the two groups (so that $\overline{X}_S - \overline{X}_{NS} = 0$) or age would not be related to blood pressure (so that $\beta = 0$): in both cases the bias would be 0.

As an example of *Case B,* let us consider sex as a confounding factor. If the difference in mean blood pressures for smokers vs. nonsmokers is the same for males and females, this difference may be regarded as the treatment effect (again assuming that no factors, other than smoking and sex, affect blood pressure). But if males have higher blood pressures than females and if males are more likely to smoke than females, the overall difference in average blood pressure between smokers and nonsmokers is biased as in Case A. Another example of Case B is Example 2.1.

To illustrate *Case C,* where the outcome is categorical and the confounding is numerical, let us suppose that we are interested in the effect of smoking on mortality, and once again we will consider age as a confounding factor. Assume the same age distributions as in the example for Case A (see Figure 3.3). Now consider, for instance, the smoking group: to each level of age corresponds a death rate, and a plot of death rate vs. age may suggest a simple relationship between them; similarly in the nonsmoking group. For instance, in Figure 3.6, we have assumed that the relationship between death rate and age could be described by an exponential curve in each group, or equivalently that the relationship between the logarithm of the death rate and age could be described by a straight line in each group.

As can be seen in Figure 3.6*b,* we have also assumed that the distance between the straight lines is the same for each age (i.e., the difference in the logarithm of the death rates is a constant $\alpha'_S - \alpha'_{NS}$). Note that this difference is the logarithm of the relative risk, since the relative risk is the ratio of the death rate in the smoking group, r_S, to the death rate in the nonsmoking group, r_{NS}. That is,

$$\log r_S - \log r_{NS} = \alpha'_S - \alpha'_{NS},$$

which implies that

$$\log \frac{r_S}{r_{NS}} = \alpha'_S - \alpha'_{NS}.$$

So we are considering a model with the same relative risk at each age. The brackets in Figure 3.6 indicate the ranges of the risks of death for smokers and nonsmokers corresponding to the age ranges of Figure 3.3. A crude relative risk obtained by dividing the overall smoker death rate by the overall nonsmoker death rate would overestimate the true relative risk, because smokers tend to be older than nonsmokers.

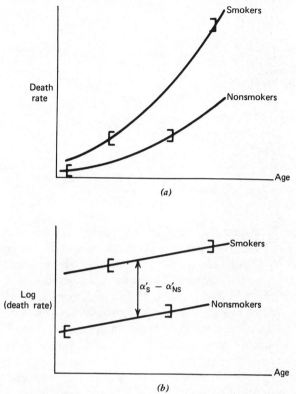

Figure 3.6 (a) *Relationship of death rate with age;* (b) *relationship of log (death rate) with age.*

Confounding in *Case D* operates much the same way as in Case B except that the initial assumption is that the relative risk of death for smokers vs. nonsmokers is the same for males and females. Example 1.1 is of the Case D type.

3.3 TREATMENT EFFECT DEPENDENT ON A BACKGROUND FACTOR

In the previous examples we have assumed an identical treatment effect for all individuals. In Figure 3.4, for instance, smoking increases blood pressure by the same amount for everybody. The assumption of constant treatment effect is commonly made for simplicity, but it may be more realistic to assume that a treatment acts differentially across individuals. This variability may be modeled by assuming that the treatment effect is a function of one or several background factors. For instance, the effect of surgery as compared with standard medication

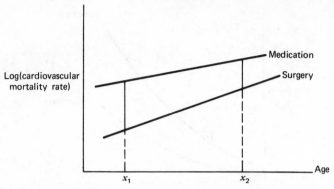

Figure 3.7.   *First example of interaction.*

in the treatment of cardiovascular diseases depends in particular on a patient's age, arterial state, and properties of the heart as measured by several variables. It may or may not be desirable to refer then to a summary measure of treatment effect, as the following two hypothetical cases will illustrate.

For simplicity we will assume that the effect of surgery depends only on age and that the relationships between age and cardiovascular mortality for both surgical and medical treatments are as shown in Figures 3.7 and 3.8; in both cases, the logarithm of the cardiovascular mortality rate is a linear function of age under each treatment. In Figure 3.8, but not in Figure 3.7, the two lines cross. In both cases, the comparison of surgery and medication depends on age. In Figure 3.7, surgery is always associated with a lower mortality rate, its greatest benefit being for younger patients (x_1). In this case, a summary treatment effect such as a difference in the average logarithm of the mortality rates would provide useful information on the effect of surgery. Contrast this with Figure 3.8, where surgery is beneficial for younger patients (x_1), whereas for older patients (x_2)

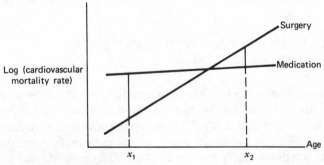

Figure 3.8   *Second example of interaction.*

standard medication is preferable. Here a summary measure would give a distorted picture of the effect of surgery.

When the treatment effect is related to a background factor in this way, there is said to be an *interaction* between the treatment and background factors. The presence or absence of interaction may depend on the measure chosen to express the treatment effect.

Example 3.1   Treatment for breast cancer: Consider the data given in Table 3.2, which come from a randomized study (Atkins et al., 1972) comparing two forms of surgical treatment for breast cancer. The outcome variable is the presence or absence

Table 3.2   *Surgical Treatment for Breast Cancer*[a]

	Surgical Procedure	
Clinical stage 1	Extended Tylectomy	Radical Mastectomy
Recurrence	15	4
No recurrence	$\dfrac{97}{112}$	$\dfrac{104}{108}$

$$\text{Rates difference} = \frac{15}{112} - \frac{4}{108} = 0.10$$

$$\text{Relative risk} = \frac{15}{112} \Big/ \frac{4}{108} = 3.62$$

$$\text{Odds ratio} = \frac{15 \times 104}{4 \times 97} = 4.02$$

	Extended Tylectomy	Radical Mastectomy
Clinical stage 2		
Recurrence	30	9
No recurrence	$\dfrac{40}{70}$	$\dfrac{71}{80}$

$$\text{Rates difference} = \frac{30}{70} - \frac{9}{80} = 0.32$$

$$\text{Relative risk} = \frac{30}{70} \Big/ \frac{9}{80} = 3.81$$

$$\text{Odds ratio} = \frac{30 \times 71}{40 \times 9} = 5.92$$

Adapted, by permission, from Atkins et al. (1972), Tables 2 to 4.

[a] Treatment = surgical procedure; outcome = recurrence; background factor = clinical stage.

of local recurrence of malignancy after the surgery. Patients were divided into two groups, depending upon the stage of the disease prior to surgery.

Since both the risk and outcome variables are categorical, three measures of treatment effect— difference in recurrence rates, relative risk, and odds ratio—may be computed for each stage (see the calculations in Table 3.2). It turns out that the relative risk is nearly the same for stage 1 and stage 2 patients (3.62 vs. 3.81), whereas the odds ratio and difference in rates depend on the stage (4.02 vs. 5.92 and 0.10 vs. 0.32). In other words, there is an interaction if the treatment effect is expressed in terms of the latter two measures, but no interaction if it is measured by the relative risk.

Since the logarithm of the relative risk is equal to the difference of the log rates ($\log \theta = \log r_1 - \log r_2$), this is an example where an analysis in the original units (recurrence rates) show an interaction, whereas an analysis in a different scale ($\log$ — recurrence rates) does not. Often, however, interactions cannot be removed by changing the scale. If in the previous example, stage 1 patients had fewer recurrences with tylectomy than with mastectomy but the opposite had been true for stage 2 patients, there would be no way of avoiding interaction. Figure 3.8 gives another example of nonremovable interaction.

Although it is desirable to avoid interaction since a single measure can then completely describe the treatment effect, sometimes, as we discussed in Section 3.1, because one measure of treatment effect is more useful than others, this measure should be used even if it does result in interaction.

REFERENCES

Atkins, H., Hayward, J. L., Klugman, D. J., and Wayte, A. B. (1972), Treatment of Early Breast Cancer: A Report after 10 Years of a Clinical Trial, *British Medical Journal*, 423–429.

Berkson, J. (1958), Smoking and Lung Cancer: Some Observations on Two Recent Reports, *Journal of the American Statistical Association*, **53**, 28–38.

Colton, T. (1974), *Statistics in Medicine*, Boston: Little, Brown.

Daniel, D., and Wood, F. S. (1971), *Fitting Equations to Data*, New York: Wiley.

Fleiss, J. S. (1973), *Statistical Methods for Rates and Proportions*, New York: Wiley.

Hanushek, E. A., & Jackson, J. E. (1977), *Statistical Methods for Social Scientists*, New York: Academic Press.

Hill, A. B. (1965), The Environment and Disease: Association or Causation? *Proceedings of the Royal Society of Medicine*, **58**, 295–300.

MacMahon, B., and Pugh, T. F. (1970), *Epidemiology, Principles and Methods*, Boston: Little, Brown.

Mosteller, F., and Tukey, J. W. (1977), *Data Analysis and Regression: A Second Course in Statistics*, Reading, MA: Addison-Wesley.

Sheps, M. C. (1959), an Examination of Some Methods of Comparing Rates or Proportions, *Biometrics*, **15**, 87–97.

Tufte, E. R. (1974), *Data Analysis for Politics and Policy*, Englewood Cliffs, N.J.: Prentice-Hall.

CHAPTER 4

Randomized and Nonrandomized Studies

Estimating a treatment effect requires the construction of a standard of comparison. As we have seen in Chapter 1, this involves a comparison group which does not receive the treatment of interest. In this chapter we will explore several ways of establishing such a comparison group, emphasizing the difference between randomization and other methods. It will be seen that a randomized allocation of subjects to a treatment and control group generally ensures that the latter is an adequate standard of comparison for the former.

We will start by defining randomization and discussing the properties that make this method particularly attractive. We will then give reasons for doing nonrandomized studies, and distinguish the different types of studies involving a comparison group. For simplicity of presentation, this chapter will be confined mainly to studies with a dichotomous risk factor.

4.1 DEFINITION OF RANDOMIZATION

Randomization is a method whereby subjects are allocated to one of the two risk factor groups by a random mechanism which assures that each individual has an equal chance of being assigned to either group. Tossing a fair coin and allocating an individual to one group or the other based on the appearance of "heads" or the use of a table of random numbers are examples of such processes. For instance, in studying the efficacy of a new medication relative to a standard one, the names of the patients could be entered sequentially, line by line, in a book, and a number from a table of random numbers could be assigned to each line. The patients assigned even numbers would be allocated to the new medication, those with odd numbers to a standard one. A more sophisticated randomized design should be used if we require equal numbers of patients in the two medication groups. After the randomization process has determined that a particular subject should be assigned to a particular group, the investigator must have enough control to implement that assignment. There is clearly no way to conduct a randomized study if the investigator must accept the assignment of people to treatment or comparison groups as determined by nature or by some institutional process (some examples will be given in Sections 4.4 and 4.5).

The primary virtue of randomization is that with high probability the two groups will be similar. Indeed, the only initial systematic difference between the two groups will be that one received the treatment and the other did not. Therefore, if the treatment has no effect, the distribution of the outcome variable in the two groups would be quite similar. In the next section we provide a more extensive discussion of the properties of randomization.

Although randomization offers important advantages, the investigator may sometimes want to consider nonrandom allocations. For example, it is possible to use systematic processes such as allocating every second subject or all the subjects with odd birth years to one of the two groups. Such processes may be much easier to administer than is randomization, and generally these systematic processes will be essentially equivalent to randomization. However, there is always a risk that the characteristic on which the allocation is done (order of arrival, birth year) is related to the outcome under study, so that its effect cannot be disentangled from that of the treatment. Haphazard processes, where no well-defined method is used to form the groups, are even more dangerous since the investigator may allocate, often unconsciously, a particular type of subject to one of the groups. Thus when assignment of subjects to treatments is under the control of the investigator, it is safest to use a random mechanism.

4.2 PROPERTIES OF RANDOMIZATION

1. *Randomization generally implies equal distribution of subject charac-*

teristics in each group and thereby facilitates causal inference. If the number of subjects in a randomized study is large, it is unlikely that the two groups differ with respect to any characteristic that can affect the outcome under study, whether or not these characteristics are known to the investigators. To illustrate this property, we consider a hypothetical study to determine whether drug X is effective, as compared to drug Y, in reducing blood pressure for patients with hypertension. The investigators identify a number of hypertensive patients. They randomize the patients into a drug X group and a drug Y group. What has been gained by randomization here?

By employing randomization, the investigators assure themselves that the groups are likely to have similar distributions of variables which can affect blood pressure. More precisely, the probability is small that potentially confounding variables differ in the two groups by a large amount. If the drug X group subsequently exhibits a substantially lower average blood pressure than does the drug Y group, randomization makes it unlikely that the difference is caused by a factor other than the drug. For example, it is unlikely that the drug X group has lower blood pressure because it consists of younger people.

2. *Randomization eliminates selection effects.* If individuals found eligible for a study are randomized into groups, there is no possibility that the investigators' initial biases or preferences about which subjects should receive what program could influence the results. For example, in the blood pressure illustration, the investigator, had he or she not randomized, may have tended to give the drug X to the more severe hypertensives. Thus a crude comparison of subsequent blood pressures between the two groups would not give a fair comparison of the two drugs. Or consider the Lanarkshire experiment carried out in schools to study the effect of milk on growth of children. It was criticized by Student (1931) because a loose design allowed the investigators to allocate, perhaps unconsciously, more milk to the poorer and ill-nourished children than to the well-fed children. Although the number of children participating in the study was large, this failure in the design prevented a clear inference about the effect of milk.

If the individuals are considered sequentially for admission to the study, the randomization scheme should be kept secret from the investigator. Otherwise, a medical researcher who knows that the next patient arriving at the hospital will be assigned by the randomization scheme to drug Y may declare that patient ineligible for the study if he or she would have favored drug X for this patient. In this case, randomization together with "blindness" of the investigator will eliminate any selection effects.

The investigator may or may not be conscious of his or her own selectivity in a nonrandomized allocation. Randomization will assure him or her as well as others that subtle selection effects have not operated. This element of persuasiveness is a definite strength of randomized studies.

Selection effects may be created in nonrandomized studies not only by the investigators, as we have just seen, but by the subjects themselves. With randomization, the subjects cannot select or influence the selection of their own treatment. Self-selection may be particularly troublesome in nonrandomized studies, since it is often difficult to isolate or to measure the variables that distinguish people who select one treatment rather than another, and hence to disentangle the treatment effect from the selection effect. Yerushalmy (1972), in studying the relationship between smoking during pregnancy and birth weight of the infant, argues that the observed difference in birth weights between the smoking and nonsmoking groups may be due to the smoker and not the smoking (i.e., that the smoking may be considered as an index characterizing some other unmeasured differences between smokers and nonsmokers). Even if we know how to measure these differences, it may be impossible to adjust for their effect (see Section 5.6). No such problems arise when randomization is employed.

3. *Randomization provides a basis for statistical inference.* The process of randomization allows us to assign probabilities to observed differences in outcome under the assumption that the treatment has no effect and to perform significance tests (Fisher, 1925). If the significance level attached to an observed difference is very small, it is unlikely that the difference is due only to chance. The purpose of a significance test is to rule out the random explanation. If it is used in conjunction with randomization, it rules out every explanation other than the treatment.

4.3 FURTHER POINTS ON RANDOMIZATION

1. *Background variables in randomized studies.* Although the primary virtue of randomization is to tend to balance the two groups with respect to background variables, it does not exclude the possibility of imbalance with respect to one or more individual characteristics. The larger the size of the groups, the less likely this possibility is; however, the investigator should make some basic checks on his or her data to verify that such an unlikely event has not happened. These checks involve comparing the distribution of background variables in the two groups, primarily those background factors which may have an important effect on the outcome factor. If the investigator finds differences between the two groups, he or she should use one of the adjustment techniques described in this book.

The University Group Diabetes Program (UGDP, 1970) provides an example of a carefully randomized study with an extensive check of possible inequalities between treatment groups. The study revealed a higher cardiovascular mortality among patients taking tolbutamide—a drug for the treatment of diabetes, until then regarded as safe—than among patients on other drugs for diabetes or a placebo. One of the controversies that emerged from this study concerned the excessive cardiovascular mortality—12.7% in the tolbutamide group as compared to 6.2% in the placebo group. Could it be explained by an unlucky randomization that happened to assign healthier patients to the control group? A committee appointed by NIH to review the available evidence (Committee for the Assessment of Biometric Aspects of Controlled Trials of Hypoglycemic Agents, 1975) confirmed that the random process had indeed allocated healthier patients to the control group. After adjusting for this particular problem, however, the committee concluded that there still existed excessive mortality in the tolbutamide group. Cornfield (1971) in his reassessment of the study, points out that when differences in background variables are observed after random allocation, the randomization scheme should be carefully reviewed to eliminate the possibility that it has been violated.

Often, before the randomization is carried out, certain factors are thought to have an important effect on the outcome. It is then advisable to form groups of individuals who are homogeneous in these factors and use randomization within these groups: this process is known as *stratified randomization*. It constitutes an insurance against differences in the distribution of major variables and reduces the random variability. It does require, however, more extensive bookkeeping, to perform the random allocation, and a more complex analysis, to take these groupings into account.

The question of designing such randomized studies and more complex types will not be discussed in this book. The interested reader is referred to texts on experimental design: see Cox (1958) for a nonmathematical presentation and Zelen (1974) and Pocock (1979) for reviews of designs in clinical trials. Kempthorne (1952) and Cochran and Cox (1957) present experimental designs for comparative studies and their analysis at a higher level.

2. *Randomization in small samples.* When the number of individuals in the study is small, the probability of imbalance on important background factors between the groups may be substantial. Precautions should be taken at the design stage to reduce this probability and to decrease the random variation around the difference in outcome. References given in the previous section should be consulted. Also relevant to studies where the individuals are considered sequentially for eligibility—patients entering a hospital, inmates arriving at a prison—are the new type of "biased coin designs" (Efron, 1971; Pocock and Simon, 1975; Simon, 1979). These designs attempt to achieve balance with respect to important background factors while preventing the investigator from being aware of which treatment group the next individual will be assigned to.

4.4   REASONS FOR THE USE OF NONRANDOMIZED STUDIES

We have outlined in Section 4.2 three well-known advantages of randomization: (*a*) it tends to balance subject characteristics between the groups and facilitate causal inference, (*b*) it eliminates selection effects, and (*c*) it provides a basis for statistical inference. Why, then, should standards of comparison be constructed in any other way? We present next some possible reasons for constructing standards of comparison by some other procedure.

1. *Nonrandomized studies are sometimes the only ethical way to conduct an investigation.* If the treatment is potentially harmful, it is generally unethical for an investigator to assign people to this treatment. An example of this is a study of the effects of malnutrition, where we simply cannot assign subjects to intolerable diets. Thus we compare malnourished populations with those on adequate diets.

2. *Nonrandomized studies are sometimes the only ones possible.* Certain investigations require the implementation of treatments that may affect people's lives. In a democratic society randomized implementation of such treatments is not always feasible. Consider, for example, the question of fluoridating a town's water supply. Let us assume that the voters in any town, or their elected representatives, have the final say about whether the water supply is fluoridated. No experimenter can make this decision. We would then have a series of towns, some of which have elected fluoridation and others which have not. The dental experience of the children in these towns can provide a great deal of useful information if properly analyzed.

3. *Nonrandomized studies are usually less expensive.* An advantage of nonrandomized studies is that they usually cost less per subject and may not require the extensive planning and control that are needed for randomized studies. This makes nonrandomized studies particularly attractive in the early stages of any research effort. Preliminary estimates of the relative importance of many background variables and their variation may be developed at a reasonable cost. These data may be important in designing future randomized experiments.

Also, if the investigator is expecting or is interested only in very large, "slam-bang" effects (Gilbert et al., 1975), nonrandomized studies may detect such differences adequately. For instance, the effect of penicillin on mortality was so obvious when it was first used that no randomized study was necessary. However, Gilbert et al. (1975, 1977), after reviewing a large number of innovations in social and medical areas evaluated by randomized and nonrandomized studies, conclude that such slam-bang effects are exceptional.

4. *Nonrandomized studies may be closer to real-life situations.* To the extent that randomization differs from natural selection mechanisms, the conditions of a randomized study might be quite different from those in which the treatment would ordinarily be applied. For example, a program may be very successful for those who choose it themselves on the basis of a media publicity campaign but ineffective when administered as a social experiment. It would then be dif-

ficult to disentangle the effect of the program from that of the experimental conditions and to generalize the results of this particular study to a natural, nonexperimental setting. Although we do not discuss this problem of "external validity," because it is primarily subject-matter-related rather than statistical, it might preclude experiments whose conditions of application are too artificial.

4.5 TYPES OF COMPARATIVE STUDIES

The investigator who does not have control over the assignment of treatment to individuals can often take advantage of situations created by nature or society. Suppose that we want to study the relation between cigarette smoking (risk factor) and lung cancer (outcome). Although we cannot randomly assign subjects to levels of the risk factor, we can still observe over time groups of people who smoke and who do not and compare the proportions of individuals who develop lung cancer in each group. This approach, called a *cohort study,* may require the observation of a large number of people if the outcome under study is rare in order to get enough "positive outcome" subjects (with lung cancer in this case). In cases of rare outcome, a more economical approach, the *case-control study,* may be considered. One would assemble a group of people with lung cancer and a group without and compare the proportions of smokers in each group.

These two designs (Cochran, 1965; WHO, 1972) can be viewed as different methods of sampling from a given population (we will later refer to that population as the "target population," i.e., the collection of individuals to whom we would like to apply the results of the study). In cohort studies, we focus on risk factor groups and take samples of exposed and unexposed subjects (smokers/ nonsmokers); in case-control studies, we focus on outcome groups and take samples of cases and noncases (with lung cancer/without lung cancer). To clarify this point, we can look at a 2 × 2 table (Table 4.1) which gives the number of subjects in the target population falling in each category. In a cohort study, we would take samples from the smoking group $(A + B)$ and the nonsmoking group $(C + D)$. In a case-control study, we would take samples from the group with lung cancer $(A + C)$ and the group without lung cancer $(B + D)$.

Table 4.1   Distribution of Target Population

	Smokers	Nonsmokers	Total
With lung cancer	A	C	$A + C$
Without lung cancer	B	D	$B + D$
Total	$A + B$	$C + D$	$A + B + C + D$

In the following discussion, we will point out arguments for and against each approach—cohort or case-control study. A detailed presentation of these two types of studies may be found in MacMahon and Pugh (1970).

4.5.1 Cohort Studies

In a cohort study, persons are selected on the basis of their exposure (or lack of exposure) to the risk factor. The outcome is measured in the subjects of each group after their selection for study.

Cohort studies may be either prospective or retrospective. If the outcome has occurred prior to the start of the study, it is a *retrospective* cohort study. If the outcome has not occurred at the beginning of the study, it is *prospective*. Retrospective studies are particularly useful when the time lag between exposure to the risk factor and the outcome is large, because the time needed to complete a retrospective study is only that needed to assemble and analyze the data. In a prospective study, for instance of smoking and lung cancer, one may have to wait 20 years or more until the risk factor has an effect. The possibility of doing a retrospective cohort study depends on the availability and reliability of records on both the risk and outcome factor. In a prospective study, the investigator can plan and control the collection of data and therefore avoid, or at least be aware of, defects in the collection of data. Note that a randomized study is a special type of prospective cohort study. Consequently, some of the problems mentioned later in this section apply also to randomized studies.

Cohort studies are preferred to case-control studies when the risk factor is rare in the target population. For example, suppose that we want to study the relation between working in a textile mill and lung cancer among all Americans. As this occupation is rather uncommon, it would not be efficient to use a case-control approach, because by selecting individuals with and without lung cancer we would find too small a proportion (if not nonexistent) of cotton textile mill workers to draw any conclusion. (However, if the study were to be restricted to a town with a high proportion of cotton workers, a case-control study might well be appropriate.) But with a cohort study, we could build a sample of these workers which is of reasonable size. Conversely, cohort studies do require very large sample sizes if the outcome is rare.

As pointed out earlier, the latent period between the exposure to the risk factor and the outcome may be very long, so that people may be lost before the outcome is measured. This may happen for several reasons: people move to another region; people do not want to participate any more (e.g., if the study requires periodic measurements); people die (and death is not the outcome under study); and so on. An analysis for handling losses to follow-up is presented in Chapter 11. This method of analysis takes care of situations in which the probability of loss is related to the risk factor. Imagine, for instance, that smokers tend to move more than nonsmokers and thus the loss rate would be higher in the exposed group than in the nonexposed group. Or people may have an incentive to participate if they belong to one of the risk factor groups: for instance, a tastier diet may be part of a treatment and encourage people to continue to participate; those in the "dull" diet may drop out more easily. More difficulties arise if the probability of loss is related to the outcome factor, as discussed in Chapter 11.

The problem of observation bias in ascertainment of the outcome is often present in cohort studies. If the observer who measures the outcome is aware of the risk factor group to which the subject belongs, he or she may be tempted to set systematically the doubtful cases in one outcome category for the treatment group and in another category for the comparison group. This kind of bias may be avoided by using a "blind" procedure wherein the observer does not know the risk factor group to which the subject belongs. Similarly, it is sometimes possible to keep the subject from knowing to what risk factor group he or she belongs, to avoid differential rates of reporting: the patient who receives a new drug may be influenced by expectations in reporting results of the medication. When both the observer and the subject are kept "blind," the study is called a "double-blind" study.

4.5.2 Case-Control Studies

In a case-control study, we assemble groups of subjects on the basis of their outcomes and then collect data on their past exposure to the risk factor. Consider, for example, the following study on the characteristics of adult pedestrians fatally injured by motor vehicles in Manhattan (Haddon et al., 1961). The investigators of this case-control study assembled 50 pedestrians fatally injured by motor vehicles to form the case group and 200 live pedestrians to form the control group. They were interested in the association of different risk factors—age, heavy drinking, and so on—with pedestrian deaths. To ensure some comparability between the two groups, they forced comparability on background variables (other than the risk factors) that they thought related to both the outcome and the risk factors, by assigning to each case four controls with the same sex, found at the same accident site, at the same time of day, and on the same day of week of accident. That is, they "matched" (see Chapter 6) cases and controls on four background variables. Then they compared the age distribution in both groups, the blood alcohol content, and other risk factor distributions in the two groups.

By using a case-control approach, the investigators were able to look at the influence of several risk factors by means of a single study. Their study was rather economical since they needed to assemble data on only 250 people; if they had used a cohort study, tens of thousands of people would have been needed to get a few pedestrian deaths in each risk factor group.

As with all nonrandomized studies, there can be no assurance of comparability between the two groups. The investigators have matched cases and controls on four background factors, but an undetected confounding factor could induce different distributions of risk factors in the two groups even if the risk factors had no effect.

A selection bias may be encountered in case-control studies when cases (or controls) have a different chance of being selected from the target population as exposed than as nonexposed. This results from the fact that the individuals participating in the study are selected after both the exposure and the outcome, so that a combined effect of the risk and outcome may influence the selection. The results of a case-control study have been published (Boston Collaborative Drug Surveillance Program, 1972) showing that the risk of myocardial infarction (MI), more commonly known as heart attack, was twice as large for heavy coffee drinkers as for others. The possibility of a selection bias was later suggested (Maugh, 1973). The cases were selected among hospitalized MI patients and the controls were patients from the same hospitals suffering from other diseases. However, the cases represent a "lucky" fraction of the MI population, since 60% of MI victims die before reaching the hospital. So the data would be consistent with an interpretation that coffee drinking has no adverse effect on the incidence of MI, but on the contrary increases the chance of survival after MI! If that were the case, the proportion of coffee drinkers would be higher among survivors of MI than among MI victims in the target population. Then the cases being sampled from the survivors of MI, rather than from the MI victims in the target population, would have a higher chance of being selected as coffee drinkers than as noncoffee drinkers. We would then find an excess of coffee drinkers among the cases, even though coffee drinking may have no effect on the incidence of MI.

Selection bias makes it hard to generalize the results of a case-control study to a target population, because it may be impossible to know how original groups of exposed and unexposed subjects were reduced by death or migration before they appear as cases or controls (Dorn, 1959; Feinstein, 1973). Important groups of subjects may never appear for observation in a case-control study, such as the 60% of MI victims who die before reaching a hospital.

Observation bias may arise in ascertaining exposure to the risk factor. If it is done by interviews, the quality of memory may be different among cases and controls. For instance, if mothers who give birth to abnormal children are interviewed about a possible exposure to X-rays during their pregnancy, they may remember better all the events that occurred during their pregnancy than will mothers who gave birth to normal children. Also, the interviewer may be inclined to get more accurate information among cases than among controls.

The estimation of effects raises special problems in case-control studies. To illustrate these difficulties, again consider Table 4.1, which classifies subjects

Table 4.2   Distribution of Individuals in a Hypothetical Case-Control Study

	Smokers	Nonsmokers	Total
With lung cancer	560	440	1000
Without lung cancer	360	640	1000

Table 4.3   Distribution of Individuals in Another Hypothetical Case-Control Study

	Smokers	Nonsmokers	Total
With lung cancer	5,600	4,400	10,000
Without lung cancer	360	640	1,000

in the target population according to their smoking and lung cancer statuses. In a case-control study we may decide to take 1000 subjects from the population with lung cancer and 1000 subjects without and look into their smoking history: we may get Table 4.2. Or we may want to choose more lung cancer patients, say 10,000, and get Table 4.3.

In Chapter 3 we discussed three measures of treatment effect available when both the risk and outcome factors are dichotomous. Applied to our example, they would be:

- The difference in lung cancer rates in the smoking and nonsmoking groups.
- The relative risks of developing lung cancer for smokers as compared to nonsmokers.
- The odds ratio, that is, the ratio of the odds of developing lung cancer in the smoking group to that of developing lung cancer in the nonsmoking group.

By comparing Tables 4.2 and 4.3, we see that the lung cancer rate in, for instance, the smoking group, is not meaningful since the number of lung cancer patients may be changed at will by the investigator: in Table 4.2, 560 of 920 (= 560 + 360) smokers have lung cancer; in Table 4.3, 5600 of 5960 (= 5600 + 360) smokers have lung cancer. Thus a measure derived from comparing lung cancer rates, such as the difference of rates and relative risk, cannot be interpreted in a case-control study. Only the odds ratio can be computed, since it does not depend on the sampling ratio of lung cancer to noncancer subjects (it is equal to 2.26 for the data in Tables 4.2 and 4.3).

Under the following two special circumstances, however, the relative risk in the target population may be estimated from the odds ratio of a case-control study:

1. The outcome is rare (this implies that the relative risk of the target population is approximately equal to the odds ratio of the target population).

2. There is no selection bias (then the odds ratio of the case-control study is a good estimate of the odds ratio in the target population). Detailed calculations showing how the relative risk in the target population may be estimated by the odds ratio are given in Appendix 4A.

Further information regarding case-control studies can be found in the proceedings of a recent symposium (Ibrahim, 1979).

4.5.3   Cross-Sectional Studies

Up to now we have presented two types of studies which are generally *longitudinal;* that is, there is a period between the exposure to the risk factor and the outcome which is the period needed by the risk factor to have an effect, if it has any. However, the length of this period may not be known; the outcome may be undetected for a while or the exposure to the treatment may expand over many years. For these various reasons, *cross-sectional* studies are sometimes done in which the risk and outcome factors are ascertained at the same time. For example, to study the relationship between obesity and heart disease, we might collect data that classified people as obese or nonobese in 1978 and with or

Table 4.4   Types of Comparative Studies[a]

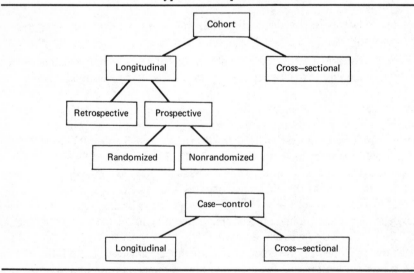

[a] All studies are nonrandomized studies except when otherwise indicated. Case-control studies can be carried out only retrospectively.

without heart disease in 1978. In addition to the usual problems in nonrandomized studies there is the difficulty of deciding whether the "outcome" or "risk" factor came first. There is no way in the previous example of indicting obesity as a causal factor in heart disease: we can imagine a circumstance where people with heart disease began to worry about their condition and ate progressively more as their disease worsened.

The different classifications we have made in this chapter are summarized in Table 4.4.

Whereas cohort and case-control studies always involve two different groups to be compared (treated and nontreated for cohort studies, cases and controls for case-control studies), there are instances in which only one group is considered, the group acting as its own comparison group: each subject is measured before and after the treatment, the first measurements providing the comparison, group, the second the treated group. The difficulty of disentangling effects due to the passage of time from the effect of the treatment is particularly troublesome in these studies. The analysis of this type of study is discussed in Chapter 12.

4.6   OUR ATTITUDE TOWARD NONRANDOMIZED STUDIES

The limitations of nonrandomized studies that have been discussed in this chapter lead to a question of research strategy. While nonrandomized studies are cheaper, more easily carried out, and can be done retrospectively, inferences from them are generally more suspect than are those from randomized studies. Does this mean that the investigator should discard the idea of doing a study at all if randomization is not feasible? Similarly, when reviewing the results of previous studies, should the reviewer discard all those with nonrandomized designs? We think not. Such a strategy would be extremely conservative.

As mentioned above, there are often sound reasons for considering nonrandomized studies, and much can be learned from them. To eliminate all such studies would be terribly wasteful. On the other hand, the researcher has a responsibility to report clearly all circumstances that may bear on the credibility of results. Without randomization, there are often many alternative explanations of observed results. The researcher must be able to present convincing evidence to rule out alternatives, or to provide data that allow the "consumer" to make an informed judgment.

APPENDIX 4A   THE ODDS RATIO AND THE RELATIVE RISK IN CASE–CONTROL STUDIES

We show how the relative risk in the target population may be estimated from the odds ratio in a case-control study if conditions 1 and 2 of Section 4.5.2 are satisfied.

Using the notation of Table 4.1, which gives numbers of smokers/nonsmokers with lung cancer/without lung cancer in the target population, we have the relative risk in the target population:

$$\theta = \frac{A/(A + B)}{C/(C + D)} = \frac{A(C + D)}{C(A + B)}.$$

If condition 1 of Section 4.5.2 is satisfied (i.e., lung cancer is rare), then A and C are small compared, respectively, to $A + B$ and $C + D$:

$$A + B \doteq B$$
$$C + D \doteq D,$$

so that $\theta \doteq AD/CB$ = odds ratio (ψ) in the target population.

Now denote by lowercase letters the numbers of subjects in a case-control study who represent a sample from the target population. Condition (2) (no selection bias) can be written, ignoring sampling variability, as

$\dfrac{a}{A} \doteq \dfrac{c}{C}$ (the selection of lung cancer patients for study does not depend on whether they smoked or not)

$\dfrac{b}{B} \doteq \dfrac{d}{D}$ (the selection of control subjects for study does not depend on whether they smoked or not),

so that

$$\frac{A}{C} \doteq \frac{a}{c} \quad \text{and} \quad \frac{D}{B} \doteq \frac{d}{b}.$$

Thus

$$\psi = \frac{AD}{CB} \doteq \frac{ad}{cb} = \hat{\psi} \qquad \text{(the odds ratio in the case-control study)}$$

Therefore, if conditions 1 and 2 are satisfied, $\hat{\psi}$, the odds ratio in a case-control study may be used as an estimate of θ, the relative risk in the target population.

REFERENCES

Boston Collaborative Drug Surveillance Program (1972), Coffee Drinking and Acute Myocardial Infarction, *Lancet* **2**, 1278–1281.

Cochran, W. G. (1965), The Planning of Observational Studies of Human Populations, *Journal of the Royal Statistical Society, Series A,* **128**, 234–266.

Cochran, W. G., and Cox, G. M. (1957), *Experimental Designs,* 2nd ed., New York: Wiley.

Committee for the Assessment of Biometric Aspects of Controlled Trials of Hypoglycemic Agents (1975), Report, *Journal of the American Medical Association,* **231,** 583–608.

Cornfield, J. (1971), The University Group Diabetes Program: A Further Statistical Analysis of the Mortality Findings, *Journal of the American Medical Association,* **217,** 1676–1687.

Cox, D. R. (1958), *Planning of Experiments,* New York: Wiley.

Dorn, H. F. (1959), Some Problems Arising in Prospective and Retrospective Studies of the Etiology of Disease, *The New England Journal of Medicine,* **261** (12), 571–579.

Efron, B. (1971), Forcing a Sequential Experiment to Be Balanced, *Biometrika,* **58,** 403–428.

Feinstein, A. R. (1973), The Epidemiologic Trohoc, the Ablative Risk Ratio, and "Retrospective" Research, *Clinical Pharmacology and Therapeutics,* **14,** 291–307.

Fisher, R. A. (1925), *Statistical Methods for Research Workers,* London: Oliver & Boyd.

Gilbert, J. P., Light, R. J., and Mosteller, F. (1975), Assessing Social Innovations: An Empirical Base for Policy, *in* C. A. Bennet and A. A. Lumsdaine, Eds., *Evaluation and Experiment: Some Critical Issues in Assessing Social Programs,* New York: Academic Press.

Gilbert, J. P., McPeek, B., and Mosteller, F. (1977), Progress in Surgery and Anesthesia: Benefits and Risks of Innovative Therapy, *in* J. P. Bunker, B. S. Barnes, and F. Mosteller, Eds., *Costs, Risks and Benefits of Surgery,* New York: Oxford University Press.

Haddon, W., Jr., Valien, P., McCarroll, J. R., and Umberger, C. J. (1961), A Controlled Investigation of the Characteristics of Adult Pedestrians Fatally Injured by Motor Vehicles in Manhattan, *Journal of Chronic Diseases,* **14,** 655–678. Reprinted in E. R. Tufte, Ed., *The Quantitative Analysis of Social Problems,* Reading, MA: Addison-Wesley, 1970, pp. 126–152.

Ibrahim, M. A., Ed. (1979), The Case-Control Study: Consensus and Controversy, *Journal of Chronic Diseases,* **32,** 1–144.

Kempthorne, O. (1952), *The Design and Analysis of Experiments,* New York: Wiley.

MacMahon, B., and Pugh, T. F. (1970), *Epidemiology, Principles and Methods,* Boston: Little, Brown.

Maugh, T. H. (1973), Coffee and Heart Disease: Is There a Link? *Science,* **181,** 534–535.

Pocock, S. J. (1979), Allocation of Patients to Treatment in Clinical Trials, *Biometrics,* **35,** 183–197.

Pocock, S. J., and Simon, R. (1975), Sequential Treatment Assignment with Balancing for Prognostic Factors in the Controlled Clinical Trial, *Biometrics,* **31,** 103–115.

Simon, R. (1979), Restricted Randomization Designs in Clinical Trials, *Biometrics,* **35,** 503–512.

Student (1931), The Lanarkshire Milk Experiment, *Biometrika,* **23,** 398–408.

University Group Diabetes Program (1970), A Study of the Effects of Hypoglycemic Agents on Vascular Complications in Patients with Adult-Onset Diabetes, *Diabetes,* **19** (suppl. 2), 747–783, 787–830.

World Health Organization (1972), Statistical Principles in Public Health Field Studies, *WHO Technical Report 510.*

Yerushalmy, J. (1972), Self-Selection: A Major Problem in Observational Studies, *Proceedings of the Sixth Berkeley Symposium on Mathematical Statistics and Probability,* Vol. 4, Berkeley, CA: University of California Press, pp. 329–342.

Zelen, M. (1974), The Randomization and Stratification of Patients to Clinical Trials, *Journal of Chronic Disease,* **27,** 365–375.

CHAPTER 5

Some General Considerations
in Controlling Bias

Before presenting the various methods for controlling bias, we will raise several important caveats. The basic theme of this chapter is that the validity of any statistical adjustment rests on a set of assumptions which may be difficult to verify. In addition to an understanding of technical details, judgment is required in order to apply these techniques properly. Since a certain amount of practical experience is necessary to develop good judgment, we can offer no simple formulas. However, we can point out the major problems that arise in practice and some general approaches which are helpful in dealing with them.

5.1   OMITTED CONFOUNDING VARIABLES

In order to obtain a valid estimate of the treatment effect, the analyst must be sure that the variables used for adjustment include all important confounding factors. In Chapter 2 we defined a confounding factor as a variable that has the following properties:

1. Is statistically associated with the risk factor.
2. Directly affects the outcome.

The main problem is to verify part 2 of this definition. The judgment that a particular variable exerts a direct causal influence on the outcome cannot be based on statistical considerations; it requires a logical argument or evidence from other investigations.

For example, suppppose that we are investigating the effectiveness of an educational program aimed at improving the reading ability of elementary school children. Two classes are being compared, one receiving the new program and one utilizing the standard curriculum. The children have been rated on scales indicating the level of parent education and family economic circumstances. Suppose that the class receiving the new program contains a higher proportion of poor children. Then, if poverty is thought to have a direct influence on reading ability, it can be considered a confounding factor. But poverty may be closely linked to parent education in a complex causal relationship. Although some of the effect of parent education may be attributable to economic circumstances per se, there may be an independent component related to education itself. So even if we compared two equally poor children receiving identical treatments, we would still expect differences in parents' education to result in different expected reading abilities. That is, conditional on economic status, parent education still constitutes a confounding factor.

Now it might seem that including either education or economic status as adjustment variables would be reasonable, even though using both would be better. Moreover, if there exist other unmeasured variables mediating the effects of these two variables in combination, failure to include them would not seem very serious. In randomized studies that is in fact the case. Omitting a relevant variable results in less precise estimation, but the estimate of effect is unbiased. In nonrandomized studies, however, serious problems can result.

To see more clearly the nature of these problems, let us consider a hypothetical example. Suppose that in reality there are only two confounding factors, X_1 and X_2. Tables 5.1 and 5.2 display the joint frequency distribution of X_1 and X_2 and the average outcome values given X_1 and X_2 under the treatment and control conditions. From the calculations shown in Table 5.2, it is clear that if we do not adjust at all, we will estimate the treatment effect as

$$\text{Estimate of effect} = 51.25 - 61.25 = -10.$$

However, we can see that for each possible combination of X_1 and X_2 values,

Table 5.1 Joint and Marginal Frequency Distributions of X_1 and X_2^a

		Treatment Group X_2						Comparison Group X_2		
		0	1	Total				0	1	Total
X_1	0	10	90	100		X_1	0	90	10	100
	1	90	10	100			1	10	90	100
	Total	100	100	200			Total	100	100	200

a Each factor has two levels, denoted by 0 and 1.

Table 5.2 Average Outcome Values

		Treatment Group X_2					Comparison X_2	
		0	1				0	1
X_1	0	100	50		X_1	0	100	50
	1	50	25			1	50	25

Estimated average outcomes ignoring X_1 and X_2:

Treatment group:
$$\frac{100(10) + 50(90) + 50(90) + 25(10)}{200} = 51.25$$

Comparison group:
$$\frac{100(90) + 50(10) + 50(10) + 25(90)}{200} = 61.25$$

Estimated average outcomes adjusting for X_1 only:

Treatment group: $X_1 = 0$ $\dfrac{100(10) + 50(90)}{100} = 55.0$

$X_1 = 1$ $\dfrac{50(10) + 25(90)}{100} = 27.5$

Comparison group: $X_1 = 0$ $\dfrac{100(90) + 50(10)}{100} = 95.0$

$X_1 = 1$ $\dfrac{50(10) + 25(90)}{100} = 47.5$

the average outcome is equal for the two groups. That is, the treatment really has no effect, and our estimate is therefore incorrect.

Now let us adjust this estimate, by conditioning on values of X_1 alone. This simple method for correcting bias constitutes a special case of stratification, which is described in detail in Chapter 7. From the information at the bottom of Table 5.2, we obtain

Estimate of effect = (estimate given $X_1 = 0$)$P(X_1 = 0)$

$$+ \text{ (estimate given } X_1 = 1)P(X_1 = 1)$$

$$= (-40)\frac{100}{200} + (-20)\frac{100}{200} = -30.$$

In this example, adjustment by X_1 alone has increased the bias from -10 to -30.

This example is rather extreme. In most practical situations, we can expect the effect of omitted variables after controlling for a few key, identifiable variables to be small. That is, inclusion of additional variables will not change the estimated effect enough to alter the interpretation of results.

Of course, there is never a guarantee that all important variables have been considered. It is the analyst's responsibility to present evidence that for individuals who are equal in terms of variables included, there is no variable that still satisfies conditions 1 and 2 given at the beginning of this section. More precisely, if a variable does satisfy these conditions, its marginal effect must be very small.

Note also that in our example, the marginal distributions of X_1 and X_2 are identical in the two groups. So it might appear that they are not confounding variables according to the definition given in Chapter 2. But if we apply the definition to the joint distribution, we see that it does apply to the pair (X_1, X_2), which together constitute a confounding factor.

The dilemma posed by statistical adjustments is that no matter what variables we include in the analysis (X_1 in our example), there may be an omitted variable (X_2 in our example) that together with the included variables constitutes a confounding variable. Moreover, it is not enough to demonstrate that all plausible confounding variables excluded have similar distributions across groups. As with our example, such a variable may still be important in combination with others. So the analyst must be fairly certain that no variable has been left out which mediates the causal effect of those variables included.

It is clear that judgment and experience are necessary in selecting variables. Also, close collaboration between statisticians and scientists in both the design and analysis of a study is highly desirable. The problems in selecting variables are primarily substantive and not statistical, although there are some statistical guidelines that may often prove useful.

Cochran (1965) suggests that the background variables be divided into three classes:

1. A small number of major variables for which some kind of matching or adjustment is considered essential. These are usually determined by knowledge of the specific subject matter and review of the literature.

2. Variables that *may* require matching or adjustment.

3. Variables that are believed to be unimportant or for which data are not available.

Decisions regarding the variables that fall in category 2 can be very difficult. The problem is similar to that of model specification in the context of multiple regression (see Cox and Snell, 1974; Mosteller and Tukey, 1977, Chap. 16). In regression analysis, we want to include enough relevant variables to ensure that the resulting model is a correct description of the relationship between an outcome variable and a set of input variables. A commonly used criterion for the importance of a particular variable, given a set of other variables, is the decrease in the proportion of explained variation when that variable is excluded. Since this number depends on which other variables are also included in the analysis, no unique measure of "importance" can be defined. However, by calculating this quantity for each variable in a proposed set and trying various plausible sets, it is often possible to get a sense of which variables play the most important causal roles.

In choosing variables for statistical adjustment, a similar idea can be applied. for each variable of a given set, the change in the adjustment that would result from omitting it can be calculated. By examining various possible combinations, we can sometimes get a good sense of which variables are the confounding factors. For example, suppose that one particular variable consistently makes a large difference in the estimated effect, regardless of which other variables are included, while all other variables have smaller effects that depend strongly on the composition of the whole variables set. In such a situation we would be satisfied to use only this one variable in our analysis.

While part 2 of the definition given at the beginning of this section is hardest to verify, part 1 is also important. A variable that is strongly related to outcome is confounding only of its distribution differs appreciably across the treatment groups. So before tackling the more difficult task described above, the analyst may want to reduce the number of potential factors by eliminating those variables with similar distributions across groups. However, as noted above, joint distributions as well as those of each variable separately must be considered.

In this context, an appreciable difference among treatment groups is not necessarily the same as a statistically significant difference. Significance tests place the burden of proof on the rejection of the null hypothesis. As we indicated in Chapter 2, a large difference will not be statistically significant if it has an even larger standard error. In small studies, associations between background variables and the treatment which are large enough to dictate the estimate of treatment effect may not be statistically significant. The opposite problem can occur in large studies. Weak association between treatment and background variables may be statistically significant and yet be too small to affect the estimate of treatment effect.

A systematic method for examining the joint distributions of background variables is discriminant analysis. The *discriminant function* is defined as that linear combination of the background variables which maximizes the ratio of the "between-group" component of variance to the "within-group" component. Among all linear combinations of the original variables, the discriminant is the one which best separates the two groups. A thorough discussion of discriminant analysis is given by Lachenbruch (1975).

Having obtained the discriminant function, we can examine the joint distributions of those variables which enter into it most prominently. Alternatively, we can take the discriminant function itself as a single new confounding factor. Since the discriminant will generally include small contributions from many relatively unimportant background variables, we may wish to screen out some variables at the outset.

Further discussion of the variable selection problem in the context of discriminant analysis is given by Cochran (1964). He considers whether the effect of including specific variables in the discriminant functions can be assessed from the discriminating power of those variables considered individually. Although standard statistical theory warns that it cannot, an examination of 12 well-known numerical examples from the statistical literature revealed the following:

1. Most correlations (among background variables) are positive.

2. It is usually safe to exclude from a discriminant, before computing it, a group of variables whose individual discriminatory powers are poor, except for any such variate that has negative correlations with most of the individually good discriminators.

3. The performance of the discriminant function can be predicted satisfactorily from a knowledge of the performance of the individual variables as discriminators and of the average correlation coefficient among the variables.

We close this section with a brief discussion of those variables for which data are not available (Cochran's class 3). As noted above, failure to collect data on an important confounding variable can put the results of the study in serious question, particularly when the magnitude of the estimated treatment effect is small (even if it is statistically significant). However, when the magnitude of the treatment effect is large, one can often say that, even if an important confounding factor had been overlooked, it could not have accounted for the size of the observed effect. Bross (1966, 1967) has devised a quantification of this argument which he calls the "size rule." The basic idea behind the size rule is to specify, for a given observed association between treatment and outcome, how large the associations between treatment and confounding factor and between confounding factor and outcome must be to explain away the observed treatment effect. However, as noted by McKinlay (1975), the derivation of Bross's rules requires assumptions that limit the applicability of his results.

5.2   MEASUREMENT ERROR

Many observed variables really reflect two kinds of information. In part the value of the variable is governed by some stable individual characteristic that can be expected to relate to other characteristics in a systematic way. In part, however, it is determined by "random" fluctuations related to the particular circumstances under which the observation happened to be taken. This error component can vary across measurement situations even if the individual has not changed.

Measurement error is particularly troublesome in the fields of education and psychology, where the variables studied are often scores on psychometric tests. Many extraneous factors besides stable individual differences may influence test scores. Psychometricians have developed the concept of *reliability* as a way to quantify the amount of measurement error. Loosely speaking, the reliability represents the proportion of total variation comprised by variation in the underlying true score. The higher the reliability, the more confidence we can have that something real is being measured.

However, true scores are not directly observable. So various indirect methods must be used to assess the reliability of a variable measured with error, or *fallible* variable. For example, under certain assumptions, the correlation between scores of the same test given individuals at two different points in time can be used to estimate the reliability. For our purposes, the general concepts of measurement error and reliability will suffice. The reader interested in more detail on these concepts is referred to Lord and Novick (1968). We now consider the effects of measurement error on statistical adjustments.

One way to describe the effects of measurement error is in terms of omitted confounding variables. The presence of error in the observed variable means that there exists, in effect, an additional variable (error) that ought to be included along with the observed score as a confounding factor. To see this more clearly, suppose that theres exists a dichotomous confounding variable T (for true score) which can have values 0 and 1.

The frequencies of the two possible values of T in the two groups are given in Table 5.3 together with the average outcome conditional on each T value. For simplicity, we assume that the real treatment effect is 0. Then if we knew T for each individual, we could calculate separate estimates of the treatment effect within the two groups ($T = 0$; $T = 1$). Except for sampling fluctuation, the correct value of 0 would result.

Table 5.3 Frequencies and Average Outcomes for $T = 0$ and $T = 1$

		Frequencies	
		Treatment	Comparison
T	0	100	25
	1	100	175
	Total	200	200
		Average Outcomes	
T	0	100	
	1	50	

Now assume that T cannot be observed directly, but we can measure a variable X that reflects both T and measurement error E, where

$$E = \begin{cases} 0 & \text{if } X = T \\ 1 & \text{if } X = 1 - T \end{cases}$$

Then there is a joint distribution of X and E in each treatment group. For example, consider the distribution shown in Table 5.4. The relationship among T, X, and E can be expressed as in Table 5.5, and the average outcome values are as given in Table 5.6. From Table 5.6 it is clear that if we could obtain information on E as well as that for X, the pair (X, E) would constitute a confounding factor. Using X alone corresponds to the use of X_1 in the example of Section 5.1, and E plays the role of X_2.

Table 5.4 Joint Frequency Distributions of X and E^a

		Treatment Group						Comparison Group		
		E						E		
		0	1	Total				0	1	Total
X	0	80	20	100		X	0	20	35	55
	1	80	20	100			1	140	5	145
	Total	160	40	200			Total	160	40	200

^{a} Each factor has two levels, denoted by 0 and 1.

Table 5.5   Relationship among T, X, and E

		\(E\) 0	1
\(X\)	0	T = 0	T = 1
	1	T = 1	T = 0

Table 5.6   Mean Outcome as a Function of X and E

Treatment Group				Comparison Group		
		\(E\) 0	1			\(E\) 0	1
\(X\)	0	50	100	\(X\)	0	50	100
	1	100	50		1	100	50

The main point of this section is that measurement error constitutes one special form in which an omitted confounding factor can arise. By adjusting on the basis of a fallible variable, we are ignoring the variable E, which is the discrepancy between X and the true score T. If we knew both X and E, we would know T and could adjust on it.

Of course, there is an implicit assumption here that adjustment on T would eliminate all bias. If this is not the case, the relationship between adjustment on the basis of X versus T is more complicated. The reader interested in more details is referred to Weisberg (1979).

5.3   THE REGRESSION EFFECT

Measurement error represents one very common example of omitted confounding variables. Another is the phenomenon of regression effects (see Thorndike, 1942). Mathematically, regression effects can be easily explained, but heuristic interpretations are often confusing. Rather than attempt a general exposition, we will discuss regression effects in the context of a concrete example.

Suppose that a remedial program is given to a group of children in a particular

school. The aim is to improve their reading ability. A pretest is given prior to
the intervention and a posttest just after the program. A hypothetical data set
is presented in Table 5.7. From these data we can calculate the mean score at
the two testing points.

$$\text{Pretest mean} = 10.0$$

$$\text{Posttest mean} = 13.0$$

So the children have gained 3.0 points during the course of the program. But
this 3.0 points represents the sum of a treatment effect plus any natural maturation that might have occurred anyway. In Chapter 12 we consider this particular kind of confounding in more detail. Our purpose here is simply to illustrate
how regression effects can occur.

Table 5.7 Hypothetical Data on Treatment Group to Illustrate Regression Effect[a]

Pretest Score	Posttest Score								
	8	9	10	11	12	13	14	15	16
13						1	1	1	1
12					1	1	2	1	1
11				1	2	3	3	2	1
10			1	1	3	4	3	1	1
9			1	2	3	3	2	1	
8			1	1	2	1	1		
7			1	1	1	1			
6									
5									

Reprinted, by permission from Campbell and Stanley (1966), Fig. IA, copyright 1966, American
Educational Research Association, Washington, D.C.

[a] Numbers indicate how many children received the particular combination of pretest and posttest
scores.

Instead of looking at the entire group of children, let us focus on those who
are farthest from the mean. Children with scores of 7 on the pretest receive an
average score of 11.5 on the posttest. Although they start out 3 points below the
mean, they end up only 1.5 points below the posttest mean. Those scoring 13 on
the pretest (3 points above the mean) end up with an average posttest score of
14.5, which is only 1.5 points above the mean.

In general, any group of children selected on the basis of their pretest scores
will (on the average) have posttest scores closer to the mean. This phenomenon
is known as *regression toward the mean*. It results from imperfect correlation
between pretest and posttest.

To understand the effect of regression toward the mean on methods for controlling bias, imagine that a comparison group has been selected from a nearby school. The data on these comparison children are shown in Table 5.8. For this comparison group

$$\text{Comparison pretest mean } = 8.0$$
$$\text{Comparison posttest mean } = 10.0$$

Because the groups started out at different levels (10.0 for treatment vs. 8.0 for comparison), a straightforward comparison of the posttest scores may be biased.

Table 5.8   Hypothetical Data on Comparison Group from a Different School[a]

Pretest	Posttest								
	7	8	9	10	11	12	13	14	15
13									
12									
11				1	1	1	1		
10			1	1	2	1	1		
9		1	2	3	3	2	1		
8	1	1	3	4	3	1	1		
7	1	2	3	3	2	1			
6	1	1	2	1	1				
5	1	1	1	1					

[a] Numbers indicate how many children received the particular combination of pretest and posttest scores.

One common approach in such situations is to match individuals with identical scores in the two groups. (Matching is discussed in detail in Chapter 6.) For example, we could compare the average scores of individuals with pretest scores of 7. Then, because these individuals all start out equal, we might expect the comparison to yield an unbiased estimate of the treatment effect. To see what actually happens in this situation, suppose that the true treatment effect is really zero. That is, the changes between pretest and posttest are entirely the result of natural growth.

Now consider what happens when we compare across groups. We have already seen that the 4 children scoring 7 on the pretest obtain an average of 11.5. What about the 12 children scoring 7 in the comparison school? These have an average posttest score of only 9.5. They have regressed toward the mean of their own population, which is 10.0 rather than 13.0. As a result, the estimated effect is

$$\hat{\alpha} = 11.5 - 9.5 = 2.0.$$

Because of differential regression to the mean in the two groups, a *regression effect* is generated. Even though there is no treatment effect, the treatment group appears to be doing better than the comparison group. Controlling for the pretest score in this manner does not eliminate bias completely.

We have presented this example at length because it represents the kind of explanation that is often given for biased estimates of effect after matching or statistical adjustment. Moreover, unlike our previous example, it illustrates the problem in the context of numerical confounding variables. However, the crux of the problem posed by the regression effect is simply that the variables used in carrying out the adjustment (e.g., a pretest or test on a related skill) represent an incomplete description of the differences between groups. That is, we have omitted some important confounding factors. The different joint distributions of pretest and posttest in the two groups represent another way to describe the fact that, conditional on the pretest, there still exist confounding factors that can bias the treatment comparison. Two children with identical pretest scores, but in different schools, do not have the same expectation on the posttest. For example, one school may already have a remedial reading program for younger children that tends to inflate pretest performance. Unless we include a variable that reflects the effect of this remedial program, the analysis can be seriously biased.

5.4 SPECIFYING A MATHEMATICAL MODEL

So far we have discussed possible problems relating to the variables used for adjustment, but have not focused on the particular method of analysis. For illustrative purposes, we have introduced simple forms of matching, or stratification, because they allow the basic issues to be seen clearly. However, the adequacy of statistical adjustment in an actual situation depends not only on using the correct variables, but also on applying a technique whose assumptions are valid.

Most of the methods we present in this book assume a particular mathematical form for the relationship among outcomes, risk variables, and covariates. These mathematical *models* will be discussed in detail in later chapters. Our purpose in this section is to discuss the general issue of proper model specification.

For simplicity we assume that only a single variable X is needed for adjustment. While in general the problems of model error and incomplete covariates are intertwined, we wish here to isolate the modeling problems. Let us define

$$Y = \text{outcome variable (numerical)}$$

$$\alpha_i = \text{treatment effect for individual } i$$

Now in general the treatment effect may vary across individuals and may even be systematically related to X (in Chapters 2 and 3 we discussed this issue of *interaction*). We shall assume here that the effect is the same for every subject.

Further, to highlight the main issues, we assume that there is no treatment effect; that is $\alpha_i = 0$ for all individuals. Figure 5.1 is an illustration of a typical relationship between Y and X in this situation. Then there exists some mathematical function, $g(X)$, relating X and the expected value of Y. That is, the average value of Y is given by $g(X)$. Now, in general, this function may differ in the two treatment groups. But with no treatment effect this would mean that the groups differed on some additional factor besides X. So because we are assuming X to be the only confounding variable, this function must be the same in both groups.

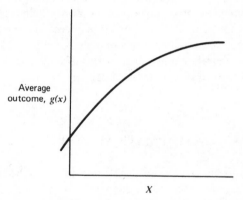

Figure 5.1   *Typical relationship between average outcome and confounding factor.*

Now, $g(X)$ is itself a variable. Moreover, because the distribution of X can differ in the two treatment groups, so can the distribution of $g(X)$. Let

$$E_1[g(X)] = \text{expected value of } g(X) \text{ in treatment group}$$
$$E_0[g(X)] = \text{expected value of } g(X) \text{ in comparison group}$$

Then we can define the bias in estimating the (zero) treatment effect as

$$E(\overline{Y}_1 - \overline{Y}_0) = E_1[g(X)] - E_0[g(X)] = \eta.$$

That is, on the average, the difference between the group means depends on the distributions of X in the two groups *and* the functional form of the relationship between outcome and covariate.

Because X can be measured in the two groups, its distribution can be determined. So if the mathematical form of g can be specified, the amount of bias can be estimated and subtracted from the raw mean difference. How, then, might this function be found?

Recall that so far we have been assuming that the treatment effect is zero, so that the same functional form holds for both groups. Now, let us suppose that the treatment has an unknown effect we wish to estimate but that it is constant across individuals. Then

$$Y = g(X) \qquad \text{for comparison group}$$
$$Y = \alpha + g(X) \qquad \text{for treatment group}$$

and

$$E(\overline{Y}_1 - \overline{Y}_0) = \alpha + \eta.$$

So we want to divide the total mean difference into two components, a part (α) attributable to the treatment and a part (η) resulting from differences between groups on the distribution of X. There are two possible ways to accomplish this: (a) use the comparison group data only to estimate g and then calculate α, and (b) fit a model including both $g(X)$ and α directly, using all the data on both groups.

Although in general it is possible to estimate any functional form, there is one class of mathematical functions that is particularly convenient: the linear functions. With only one X, a linear relationship has the form

$$g(X) = \mu + \beta X.$$

The graph of such a function is a straight line. A useful property of linear functions is that the average value of a function of X is the function of the average value of X. This means that

$$E(\overline{Y}_1 - \overline{Y}_0) = \alpha + \mu + \beta \overline{X}_1 - (\mu + \beta \overline{X}_0)$$
$$= \alpha + \beta(\overline{X}_1 - \overline{X}_0) = \alpha + \eta$$

that is,

$$\eta = \beta(\overline{X}_1 - \overline{X}_0).$$

Therefore, we can form the estimate

$$\hat{\alpha} = \overline{Y}_1 - \overline{Y}_0 - \beta(\overline{X}_1 - \overline{X}_0),$$

and we will have

$$E(\hat{\alpha}) = \alpha + \eta - \eta = \alpha.$$

The details of this approach will be elaborated upon in Chapter 8, where we refer to it as the analysis of covariance.

The assumption of linearity greatly facilitates the analysis of data from nonrandomized studies. This assumption is at the heart of several techniques discussed in this book. Although it may seem that linearity is a very strong condition, it still allows a certain amount of flexibility when used in conjunction with transformation of the data. Even though the relationship between Y and X may not be linear, it may be possible to rescale either or both to bring about linearity. For example, if the relationship between Y and X is exponential,

$$g(X) = e^{\mu + \beta X},$$

Then

$$\log g(X) = \mu + \beta X.$$

So by using the logarithm of Y as the outcome measure, a linear model analysis is possible.

The estimation of $g(X)$ as the basis for statistical adjustment allows a much more efficient use of the data than do such approaches as matching or stratification, which do not depend on a model. In matching, for instance, it may be difficult to find a large enough number of close matches to allow precise estimation. This issue is discussed in detail in Chapter 6. By assuming a mathematical structure, we may be able to estimate α precisely using relatively small sample sizes.

On the other hand, if the model used turns out to be incorrect, our results may be misleading. Suppose, for example, that we are using the comparison group data to estimate $g(X)$ in the absence of the treatment, and that $g(X)$ actually has the nonlinear form illustrated in Figure 5.2. The numbers at the bottom of Figure 5.2 represent the X values for comparison (0) and treatment group (1) subjects. Suppose that we estimate a linear model based on comparison group

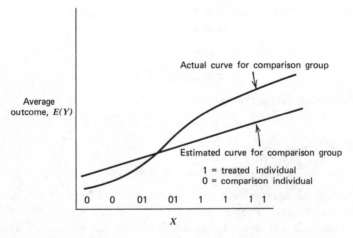

Figure 5.2   *Model misspecification.*

data. Then for individuals with high X values, the linear function underestimates their expected outcome. For those with low X values, the expected outcome is overestimated. If the treatment group tends to lie near the high end, as shown in Figure 5.2, and the control group near the low end, the actual outcome difference η produced by this difference will be much larger than that estimated on the basis of linearity. The estimate of α will be correspondingly biased.

It is hard to say how severe the departure from the assumed model must be to cause serious problems. Determining an adequate model requires judgment as well as a knowledge of particular statistical methods. In each of the subsequent chapters on individual techniques, more detail will be given on the model assumptions and how they can be verified.

Finally, we note that problems of variable selection, including measurement error and regression effects, are intertwined with those of model selection. A correctly specified model must include appropriate variables *and* have a proper mathematical form. When we transform a variable, we change both the variable and the functional form. What matters is whether the model and variables ultimately employed in the analysis accurately represent the underlying phenomenon.

5.5 SAMPLING ERROR

Throughout the previous discussion we have largely ignored the fact that analyses are often based on small or moderate sample sizes. We have focused on problems that will cause the estimated effect to deviate from the actual effect even with very large samples. We now discuss an additional source of error, that attributable to sample fluctuation.

For illustrative purposes, suppose that the true model underlying a set of data is given by the following equations:

Treatment: Average outcome $= 5 + X$

Comparison: Average outcome $= 2 + X$

This situation is illustrated in Figure 5.3. The treatment effect is 3 in this example.

Of course, in a real situation we will not know the exact relationship between X and the outcome. The problems in using the wrong mathematical model were discussed in Section 5.4. We showed, for example, that using a linear function when a nonlinear one is appropriate can lead to bias in estimating the treatment effect. Now let us assume that the functional form is in fact linear, but that we must estimate the slope and intercept from a given set of data. The sets of observed values may look as shown in Figure 5.4. The model states that *on the average,* for a given value of X, the treated individuals have a value $5 + X$ and

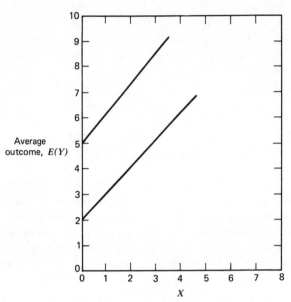

Figure 5.3 *Linear relationship between average outcome and confounding factor: constant treatment effect.*

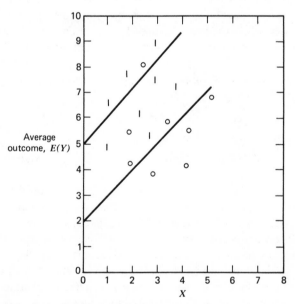

Figure 5.4 *Scatter of actual observations around expected value.*

the control subjects $2 + X$. However, the individual scores fluctuate randomly about these lines. If the number of subjects is very large, it is possible to estimate the true intercepts and slopes with great precision. However, with a sample of only 20 or 30 in each group, there can be substantial variation in these estimates from sample to sample.

Without going into detail on statistical techniques, let us imagine that we are estimating the slope using the data from the two groups combined. The difference between the estimated intercepts is then the estimator of the treatment effect. For 10 independent samples with 20 in each group, we would obtain results that vary around the true value of 3, but differ from sample to sample. The estimator may be correct on the average and therefore be what is called by statisticians an *unbiased* estimator, as mentioned in Chapter 2. However, for any particular sample there will be a *sampling error,* which may be substantial. The sampling error will generally become smaller and smaller as the sample size increases, although for some estimators it is not negligible, even for very large samples. A precise consideration of these matters would involve technicalities beyond the level of this book. Unless otherwise stated, we can assume that sampling error will disappear for a large-enough sample size.

5.6 SEPARATION OF GROUPS ON A CONFOUNDING FACTOR

In order for a confounding factor to create substantial bias in estimating a treatment effect, its distribution in the two treatment groups must differ significantly. However, if the groups are very widely separated on a confounding variable used in the analysis, certain problems mentioned in the previous sections become particularly severe. Figure 5.5 illustrates the situation where the groups are widely separated on a variable X. We have mentioned that one basic approach to bias control is the comparison of individuals with identical (or similar) values of X. This matching, if exact, will remove any bias attributable to X regardless of the functional form of the relationship between outcome and X. But it is clear that if the groups are completely separated, no matches can be found. More generally, if there is little overlap, matches may be found for only a small proportion of subjects. The feasibility of matching with different degrees of separation is discussed more completely in Chapter 6.

A second problem is that extreme separation may be an indicator that the two groups are quite different in character. So there are likely to be other variables on which they differ that are related to the outcome. It may be difficult to find a few variables that capture all the relevant variation. For example, suppose that the treatment group includes only individuals under 35 years of age and the control group contains only individuals over 35. Then the groups represent dif-

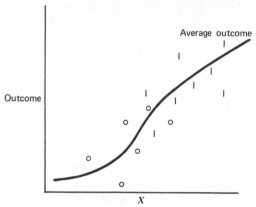

Figure 5.5 *Wide separation of groups on the confounding factor.*

ferent "generations," with quite different experiences at comparable ages and quite different life-styles. It may be meaningless to compare such groups.

A third problem is the loss of precision in estimating a model relating the expected outcome to the confounding variable. In Section 5.4 we mentioned that one approach is to estimate the function on the basis of the comparison group data only. However, when the groups are widely separated, there will be very few observations on comparison subjects in the range of X values occupied by the treatment group, and so it will be hard to obtain a precise estimate. In the case of complete separation, we must rely on extrapolation of the estimated function completely beyond the range of the data, a procedure that is always hazardous.

If we assume a known functional form and a constant treatment effect, we can, instead, estimate the treatment effect from the data on both groups. However, if there are very few observations in the range where the two distributions overlap, we must rely heavily on model assumptions, such as the assumption of no interaction. An incorrect model specification will be very difficult to detect.

The problem of complete, or near-complete, separation may sometimes arise from the desire to give a certain treatment to those who are thought to need it most. Thus there may be a conflict between research design criteria and ethical considerations. Sometimes this conflict can be resolved by an imaginatively designed study. Mather et al. (1971) report on a study of 1203 episodes of acute myocardial infarction (heart attacks). The purpose of the study was to compare home care by the family doctor with hospital treatment initially in an intensive care unit. Normally, such a comparison would be impossible—the less severely ill patients would be sent home, the emergency cases to the intensive care unit. We can imagine an index of "severity" being measured on each patient. This would clearly satisfy our definition of a confounding factor, but the distribution of this factor within the home-care and hospital-care group would not have substantial overlap.

Here, however, there was agreement between various hospitals and doctors participating in the study that while some patients would clearly need hospitalization and others should clearly be treated at home, there were some patients for whom the decision was not clear-cut. For these patients, randomization was used to decide between home care and hospitalization. The decision on acceptability of a random assignment was made by the patient's own doctor, before he knew what the result of the randomization would be.

In all, 343 cases were allocated at random, and subsequent analysis confirmed that the randomized groups did not differ substantially in composition with respect to other background variables, such as age, past history of heart disease, and blood pressure when first examined. It was found that the randomized group treated at home had a 44% lower mortality than did the randomized group treated in the hospital. As might be expected, the experience of the other two groups was very different. The conclusion that home care is better than hospitalization had only been firmly established for the randomized group, although we might speculate that it would also hold for at least some other individuals.

As a final comment, we note that when there are several potential confounding variables, it is possible that the two groups are completely separated on these variables considered jointly, although the distributions of each variable individually do have substantial overlap. Consider the situation illustrated by Figure 5.6, where X_1 and X_2 represent two background variables: X_1 = age (decades),

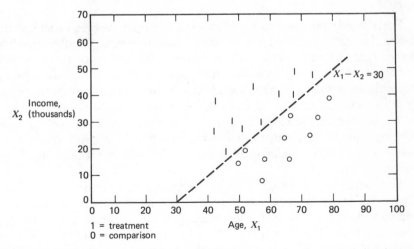

Figure 5.6 *Complete separation of joint distributions even though marginal distributions overlap.*

X_2 = income. Individuals in group 1 tend to have a higher income and lower age than those in group 0, but the distributions of each variable in the two groups have substantial overlap. However, the particular age–income combinations that occur in one group do not occur in the other. So the groups are separated on X_1 and X_2 jointly, even though they are not separated on X_1 and X_2 individually.

This problem can be very hard to recognize and emphasizes the need for a multivariate exploration of the potential confounding factors. In this example, the discriminant method might suggest that

$$D = X_1 - X_2$$

can be used to distinguish the treatment and control subjects. Note that all individuals with values of D greater than 30 are in the control group, while those with D less than 30 are in the treatment group.

5.7 SUMMARY

In this chapter we have discussed the general problems that may affect statistical adjustment strategies applied to nonrandomized studies. These problems may be seen as operating at three levels:

1. Variable selection.
2. Specifying form of mathematical model.
3. Small-sample fluctuation.

Variable selection involves knowledge of the substantive area under investigation. The aim is to include enough information to ensure that after adjusting for the measured variables, there will be no bias in the estimate of treatment (risk factor) effect. There will be no bias if the only systematic difference between two individuals with identical measured values is directly caused by the treatment (risk factor). In attempting to verify this assumption, statistical methods may be helpful, but only in conjunction with a careful analysis of possible causal relationships. Ideally, statisticians and substantive researchers should work together to select a variable set that can be defended on both statistical and conceptual grounds.

Two particularly common problems are measurement error and regression effects. We have pointed out how these can be viewed as special cases of omitted confounding factors. As such, they do not pose different problems or require special solutions. If a proposed variable set includes fallible variables, or those subject to "regression," it simply means that we must be sure to include enough other variables so that the total set is adequate.

Having an adequate set of adjustment variables allows us *in principle* to obtain an unbiased estimate of the treatment effect. However, to obtain such an estimate we must employ one of the techniques described in the subsequent chapters of this book. Each analysis strategy is based on a particular set of assumptions about the mathematical form of relationships among variables. To the extent that these assumptions do not hold in a given situation, the results may be biased.

Finally, even with an adequate set of variables and a correctly specified model, we are subject to problems arising from finite samples. That is, the estimate obtained from a particular analysis may contain a component attributable to random fluctuations. For very large sample sizes, we would expect these errors to be negligible, but for small samples we can expect the estimate to deviate substantially from the true effect. Where possible, confidence bounds should be provided in addition to a point estimate.

In each of Chapters 6 to 11 we present the basic concepts and mechanics underlying one approach to statistical adjustment. Each of these techniques is vulnerable to the general problems described in this chapter, and we will not repeat in each chapter the general caveats given here. However, we will explain in some detail how these considerations apply to the particular technique, trying to indicate what problems are most likely to arise and how to deal with them.

After reading these chapters, the reader should have a clearer understanding of the issues raised in this chapter. In Chapter 14 we will review some of these issues and present additonal areas related more specifically to the methods described in Chapters 6 to 11.

REFERENCES

Bross, I. D. J. (1966), Spurious Effects from an Extraneous Variable, *Journal of Chronic Disease,* **19,** 736-747

Bross, I. D. J. (1967), Pertinency of an Extraneous Variable, *Journal of Chronic Disease,* **20,** 487-495.

Campbell, D. T., and Stanley, J. G. (1966), *Experimental and Quasi-experimental Designs for Research,* Chicago: Rand McNally.

Cochran, W. G. (1964), On the Performance of the Linear Discriminant Function, *Technometrics,* **6,** 179-190.

Cochran, W. G. (1965), The Planning of Observational Studies of Human Populations, *Journal of the Royal Statistical Society, Series A,* **128,** 254-266.

Cox, D. R., and Snell, F. J. (1974), The Choice of Variables in Observational Studies, *Journal of the Royal Statistical Society, Series C,* **23,** 1, 51-59.

Lachenbruch, P. A. (1975), *Discriminant Analysis,* New York: Hafner.

Lord, F. M., and Novick, M. (1968), *Statistical Theories of Mental Test Scores,* Reading, MA: Addison-Wesley.

McKinlay, S. (1975), The Design and Analysis of the Observational Study—A Review, *Journal of the American Statistical Association,* **70,** 503-520.

Mather, H. G., Pearson, N. G., Read, K. L. Q., Shaw, D. B., Steed, G. R., Thorne, M. G., Jones, S., Guerrier, C. J., Eraut, C. D., McHugh, P. M., Chowdhurg, N. R., Jafary, M. H., and Wallace, T. J. (1971), Acute Myocardial Infarction: Home and Hospital Treatment, *British Medical Journal,* **3,** 334-338.

Mosteller, F., and Tukey, J. W. (1977), *Data Analysis and Regression,* Reading, MA: Addison-Wesley.

Thorndike, F. L. (1972), Regression Fallacies in the Matched Group Experiment, *Psychometrika,* **7**(2), 85-102.

Weisberg, H. I. (1979), Statistical Adjustments and Uncontrolled Studies, *Psychological Bulletin,* **86,** 1149-1164.

CHAPTER 6

Matching

The major concern in making causal inferences from comparative studies is that a proper standard of comparison be used. A proper standard of comparison (see Chapter 1) requires that the performance of the comparison group be an adequate proxy for the performance of the treatment group if they had not received the treatment. One approach to obtaining such a standard is to choose study groups that are comparable with respect to all important factors except for the specific treatment (i.e., the only difference between the two groups is the treatment). Matching attempts to achieve comparability on the important potential confounding factor(s) at the design stage of the study. This is done by appropriately selecting the study subjects to form groups which are as alike as is possible with respect to the potential confounding variable(s). Thus the goal of the matching approach is to have no relationship between the risk and the potential confounding variables in the study sample. Therefore, these potential confounding variables will not satisfy part 1 of the definition of a confounding variable given at the beginning of Chapter 2, and thereby will not be confounding variables in the final study sample. This strategy of matching is in contrast to the strategy of adjustment, which attempts to correct for differences in the two groups at the analysis stage.

We stated that matching "attempts to achieve comparability" because it is seldom possible to achieve exact comparability between the two study groups. This is especially true in the case of several confounding variables. To judge how effective the various matching procedures can be in achieving comparability and thus reducing bias in the estimate of the treatment effect, it is necessary to model the relationship between the outcome or response variable and the confounding variable(s) in the two treatment groups. Since much of the research has been done assuming a numerical outcome variable that is linearly related to the confounding variable, we will tend to emphasize this type of relationship. The reader should not believe, however, that matching is applicable only in this case. There are matching techniques which are relatively effective in achieving comparability and reducing bias in the case of nonlinear relationships.

Before presenting the various matching techniques, we shall illustrate in Section 6.1 how making the two treatment groups comparable on an important confounding variable will eliminate the bias due to that variable in the estimate of the treatment effect. Section 6.1 expands on material presented in Section 3.2.

The degree to which the two groups can be made comparable depends on (*a*) how different the distributions of the confounding variable are in the treatment and comparison groups, and (*b*) the size of the comparison population from which one samples. These factors influence the amount of bias reduction possible using any of the matching techniques, and are discussed in Section 6.2.

In the last introductory section of this chapter, Section 6.3, we list and discuss the conditions under which the results for the various matching techniques are applicable. Although these conditions are somewhat overly restrictive, they are necessary for a clear understanding of the concepts behind the various techniques.

Finally, the main emphasis of this chapter is on the reduction of the bias due to confounding. The other two sources of bias, bias due to model misspecification and estimation bias, however, can also be present. See Sections 5.4 and 5.5 for a discussion of these other sources of bias. All of the theoretical results that we present are for the case of no model misspecification. This should be kept in mind when applying the results to any study.

6.1 EFFECT OF NONCOMPARABILITY

For the sake of illustration, reconsider the example introduced in Chapter 3, the study of the association between cigarette smoking and high blood pressure. Recall that cigarette smoking is the risk variable and age is an important confounding variable. This last assumption implies that the age distributions of the smokers and nonsmokers must differ: otherwise, age would not be related to the risk variable (i.e., the groups would be comparable with respect to age). We shall further assume that the smokers are generally older (see Figure 3.3) and that the average blood pressure increases with age at the same rate for both smokers and nonsmokers (see Figure 3.4). Let X denote age in years and Y denote diastolic blood pressure in millimeters of mercury (mm Hg). The effect of the risk factor, cigarette smoking, can be measured by the difference in average blood pressure for any specific age, and because of the second assumption, this effect will be the same for all ages.

These two assumptions can be visualized in Figure 6.1. Suppose that we were to draw large random samples of smokers and nonsmokers from the populations shown in Figure 3.3. The sample frequency distributions would then be as illustrated in Figure 6.1 by the histograms. The smokers in the sample tend to be older than the nonsmokers. In particular, the mean age of the smokers is larger than that of the nonsmokers, $\overline{X}_S > \overline{X}_{NS}$. (Notice that the Y axis in Figure 6.1 does not correspond to the ordinate of the frequency distributions.)

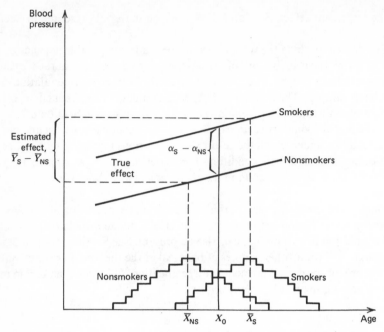

Figure 6.1 *Estimate of the treatment effect for the blood pressure–smoking example.*

The second assumption, specifying that the relationship between age and diastolic blood pressure in both groups is linear, is represented by the lines labeled "Smokers" and "Nonsmokers" (as in Figure 3.4). Algebraically, these relationships are:

$$Y_S = \alpha_S + \beta X \qquad \text{for smokers}$$

$$Y_{NS} = \alpha_{NS} + \beta X \qquad \text{for nonsmokers,} \qquad (6.1)$$

where Y_S and Y_{NS} represent the average blood pressure levels among persons of age X, and β is the rate at which Y, blood pressure, changes for each 1-year change in X. [Note that for simplicity of presentation, random fluctuations or errors (Section 2.2) will be ignored for now.] For a specified age, X_0, therefore, the effect of the risk factor is

$$Y_S - Y_{NS} = \alpha_S - \alpha_{NS} + \beta(X_0 - X_0)$$

$$= \alpha_S - \alpha_{NS} \qquad (6.2)$$

(see Figure 6.1).

Let us first consider the simplest situation, where there is only one subject in each group and where each subject is age X_0. We will then have two groups that are exactly comparable with respect to age. The estimate of the treatment effect is the difference between the blood pressures of the two subjects. Since the blood pressures of these two subjects are as given in (6.1) with $X = X_0$, the estimated treatment effect will be as given in (6.2). Thus exact comparability has led to an unbiased estimate of the treatment effect. (Note that the same result would also hold for any nonlinear relationship between X and Y.)

Next consider the estimate of the treatment effect based on all subjects in the two samples. The estimate is found by averaging over all the values of Y in both groups and calculating the difference between these averages:

$$\overline{Y}_S - \overline{Y}_{NS} = \alpha_S - \alpha_{NS} + \beta(\overline{X}_S - \overline{X}_{NS}). \tag{6.3}$$

Thus because of the noncomparability of the two groups with respect to age, the estimate of the risk effect is distorted or biased by the amount $\beta(\overline{X}_S - \overline{X}_{NS})$. Since we do not know β, we cannot adjust for this bias. (An adjustment procedure based on estimating β is analysis of covariance; see Chapter 8.) Notice, however, that if we could equalize the two sample age distributions, or in the case considered here of a linear relationship, restrict the sampling so that the two sample means were equal, we would then obtain an unbiased estimate of the treatment effect. By making the groups comparable, one would be assured of averaging over the same values of X.

There are two basic approaches to forming matches to reduce bias due to confounding. These are referred to as pair and nonpair matching. *Pair matching* methods find a specific match (comparison subject) for each treatment subject. It is clear that if we restrict the choice of subjects in the two groups such that for every treatment subject with age X_0 there is a comparison subject with exactly the same age, then by (6.2) the difference in blood pressures between each matched pair is an unbiased estimate of the treatment effect. Hence the average difference will also be unbiased.

Because of difficulties in finding comparison subjects with exactly the same value of a confounding variable as a treatment subject, various pair matching methods have been developed. For example, if the confounding variable is numerical, it is practically impossible to obtain exact matches for all treatment subjects. An alternative method, caliper matching, matches two subjects if their values of X differ by only a small tolerance (Section 6.4). In the case of a categorical confounding variable, one can use a pair matching method called stratified matching (Section 6.6). However, these methods cannot always guarantee the desired sample size, so another pair matching method, called nearest available pair matching (Section 6.5), was developed by Rubin (1973a).

In the second approach to matching, *nonpair matching,* no attempt is made to find a specific comparison subject for each treatment subject. Thus there are no identifiable pairs of subjects. There are two nonpair matching methods: frequency and mean matching. In frequency matching, Section 6.7, the distribution of the confounding variable in the treatment group is stratified and one attempts to equalize the two distributions by equalizing the number of treatment and comparison subjects in each stratum. Mean matching, Section 6.8, attempts to reduce the amount of bias by equating just the sample means rather than attempting to equalize the two distributions as in the previous methods. The comparison group, which is of the same size as the treatment group, thus consists of those subjects whose group mean is closest to the mean of the treatment group.

6.2 FACTORS INFLUENCING BIAS REDUCTION

None of the matching methods requires the fitting of a specific model for the relationship between the response and the confounding variables. The effectiveness of a matching procedure, however, will depend on the form of the relationship between the response and the confounding variables. In addition, the effectiveness depends on the following three factors: (a) the difference between the means of the treatment and comparison distributions of a confounding variable, (b) the ratio of the population variances, and (c) the size of the control sample from which the investigator forms a comparison group. These three factors will now be discussed in detail.

To understand how these three factors influence the researcher's ability to form close matches and hence to achieve the maximum bias reduction, consider the slightly exaggerated distributions of a confounding variable, X, in the treatment and comparison populations shown in Figure 6.2. Both distributions are normal with a variance of 2.25. The mean of the comparison population is 3, and the mean of the treatment population is 0.

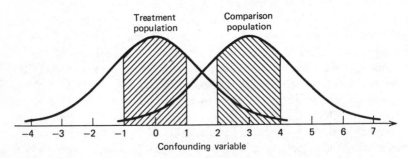

Figure 6.2 *Nonoverlapping samples, equal variances.*

Suppose that we have small random samples from both the treatment and the comparison populations and we wish to find a matched comparison group. Because of the assumed distribution, the treatment group is most likely to have values between -1 and $+1$, the middle 50% of the distribution (shaded area on the left in Figure 6.2). The sample from the comparison population is called the *comparison* or *control reservoir;* it is the group of subjects from which one finds matches for the treatment group. Based on the assumed distribution of the confounding variable, the comparison reservoir is most likely to consist of subjects whose values of the confounding variable lie between 2 and 4 (shaded area on the right in Figure 6.2). Thus there would be little overlap between these two samples.

With virtually no overlap between our samples, it is impossible to form matched groups which are comparable. Using any of the pair matching techniques, we could not expect to find many comparison subjects with values of X closer than 1 unit to any treatment subject. Similarly for the nonpair matching methods, regardless of the way one stratifies the treatment frequency distribution, there will not be enough comparison subjects in each stratum. In addition, the means of the two groups would be about 3 units apart. Any attempt to match in this situation would be unwise, since only a small proportion of the two groups could be made reasonably comparable.

Continuing this example, suppose that another, much larger sample is drawn from the comparison population such that the values of X in the reservoir lie between zero and 6. The treatment group remains fixed with values of X between $+1$ and -1. The resulting overlap of the two samples is shown in Figure 6.3 against the background of the underlying population distributions. Notice that by increasing the size of the comparison sample, we are more likely to have members of the comparison reservoir which have the same or similar values of X as members of the treatment group. The number and closeness of the possible pair matches has improved; for frequency matching we should be able to find more comparison subjects falling in the strata based on the treatment group; and the difference in the sample means, after mean matching, should be less than the previous value of 3. Again, as was the case with nonoverlapping samples, we may still be unable to find adequate matches for all treatment group subjects. This "throwing away" of unmatchable subjects is a waste of information which results in a lower precision of the estimated treatment effect.

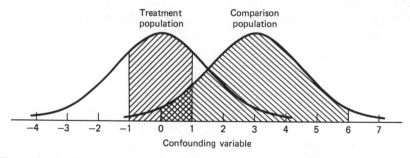

Figure 6.3 *Overlapping samples, equal variances.*

Now consider what would happen if the population variances of the confounding variable were not equal. In particular, suppose that the variance of the treatment population, σ_1^2, remains at 2.25, while the variance of the comparison population, σ_0^2, is 9.0. (Again, this is a slightly exaggerated example but is useful to illustrate our point.) With the treatment sample fixed, random sampling from the comparison population would most likely result in a sample as shown by the shading in Figure 6.4. Notice the amount of overlap that now exists between the treatment group and the comparison reservoir. There are clearly more subjects in the comparison reservoir, with values of the confounding variable between $+1$ or -1, than in the previous example (Figure 6.3).

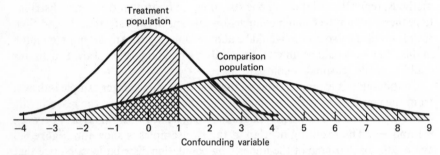

Figure 6.4 *Overlapping samples, unequal variances.*

After comparing these examples, the relationship among the three factors—the difference between the population means of the two distributions, the ratio of the population variances, and the size of the comparison reservoir—should be clear. The farther apart the two population means are, the larger the comparison reservoir must be to find close matches, unless the variances are such that the two population distributions overlap substantially.

To determine numerically the bias reduction possible for a particular matching technique, it is necessary to quantify these three factors. Cochran and Rubin (1973) chose to measure the difference between the population means by a quantity referred to as the *initial difference*. This measure, B_X, may be viewed as a standardized distance measure between two distributions and is defined as

$$B_X = \frac{\eta_1 - \eta_0}{\sqrt{(\sigma_1^2 + \sigma_0^2)/2}}. \tag{6.4}$$

The eta terms, η_1 and η_0, denote the means of the treatment and the comparison populations, respectively. Similarly, σ_1^2 and σ_0^2 represent the respective population variances.

In the first example of this section, the initial difference was equal to 2.0. With the variance of the comparison population increased to 9, however, the initial difference was equal to 1.3 and the two distributions overlapped more.

The ratio of the treatment variance to the comparison variance, σ_1^2/σ_0^2, is the second important factor in determining the number of close matches that can be formed, and hence the bias reduction possible. Generally, the smaller the ratio, the easier it will be to find close matches.

The last factor is the size of the comparison reservoir from which one finds matches. In the previous examples we assumed that the random sample from the treatment population was fixed. That is, we wanted to find a match for every subject in that sample and the subjects in the treatment group could not be changed in order to find matches. Removal of a treatment subject was the only allowable change if a suitable match could not be found. This idea of a fixed treatment group is used in the theoretical work we cite and is perhaps also the most realistic approach in determining the bias reduction possible. An alternative and less restrictive approach assumes that there exists a treatment reservoir from which a smaller group will be drawn to form the treatment group. Such an approach would allow for more flexibility in finding close matches.

In the following methodological sections the size of the comparison reservoir is stated relative to the size of the fixed treatment group. Thus a comparison reservoir of size r means that the comparison reservoir is r times larger than the treatment group. Generally, r is taken to be greater than 1.

6.3   ASSUMPTIONS

In discussing the various matching procedures, we shall make the following assumptions:

1. There is one confounding variable.
2. The risk variable in cohort studies or the outcome variable in case-control studies is dichotomous.
3. The treatment effect is constant for all values of the confounding variable. (This is the no interaction assumption of Section 3.3.)
4. For cohort studies we wish to form treatment and comparison groups of equal size. (For case-control studies, we would construct case and control groups of equal size.)
5. The treatment group (or case group) is fixed.

The assumption of only one confounding variable is made for expository purposes. In Section 6.10 we will discuss matching in the case of multiple confounding variables. The second assumption corresponds to the most common situation where matching is used. Matching cannot be used if the risk variable in cohort studies or the outcome variable in case-control studies is numerical. The third assumption of no interaction, or parallelism, is crucial for estimating the treatment effect. Researchers should always be aware that implicitly they are making this assumption and when possible they should attempt to verify it. For example, in Section 5.2, we discuss how the assumption of parallelism may be unjustified when one is dealing with fallible measurements. If this assumption is not satisfied, the researcher will have to reconsider the advisability of doing the study or else to report the study findings over the region for which the assumption holds. The fourth assumption, that the treatment and comparison groups are of equal size, is also made for expository purposes. In addition, the efficiency of matching is increased with equal sample sizes for a given total sample size. In Section 6.11 we consider the case of multiple comparison subjects per treatment subject. The last assumption of a fixed treatment group is one of the assumptions under which most of the theoretical work is done. A fixed treatment group is typically the situation in retrospective studies where the group to be studied, either case or exposed, is clearly defined.

While the type of study has no effect on the technique of matching, the forms of the outcome and confounding variables do. The various matching techniques can be used in either case-control or cohort studies. The only difference is that in a cohort study one matches the groups determined by the risk or exposure factor, whereas in a case-control study, the groups are determined by the outcome variable. Throughout this chapter, any discussion of a cohort study applies also to a case-control study, with the roles of the risk and outcome variables reversed.

The form of the risk or outcome variable and the confounding variables (i.e., numerical or categorical) determines the appropriate matching procedure and whether matching is even possible. If the confounding variable is of the unordered categorical form, such as religion, there is little difficulty in forming exact matches. We shall, therefore, make only passing reference to this type of confounding variable. Instead, we shall emphasize numerical and ordered categorical confounding variables, where the latter may be viewed as having an underlying numerical distribution. Numerical confounding variables are of particular importance because exact matching is very difficult in this situation. Most of the theoretical work concerning matching has been done for a numerical confounding variable and dichotomous risk variable (cohort study).

6.4 CALIPER MATCHING

Caliper matching is a pair matching technique that attempts to achieve comparability of the treatment and comparison groups by defining two subjects to be a match if they differ on the value of the numerical confounding variable,

X, by no more than a small tolerance, ϵ. That is, a matched pair must have the property that

$$|X_1 - X_0| \le \epsilon.$$

The subscript 1 denotes treatment group and 0 denotes comparison group. By selecting a small-enough tolerance ϵ, the bias can in principle be reduced to any desired level. However, the smaller the tolerance, the fewer matches will be possible, and in general, the larger must be the reservoir of potential comparison subjects.

Exact matching corresponds to caliper matching with a tolerance of zero. In general, though, exact matching is only possible with unordered categorical confounding variables. Sometimes, however, the number of strata in a categorical variable is so large that they must be combined into a smaller number of strata. In such cases or in the case of ordered categorical variables, the appropriate pair matching technique is stratified matching (Section 6.6).

6.4.1 Methodology

To illustrate caliper matching we shall consider the cohort study of the association of blood pressure and cigarette smoking.

Example 6.1 *Blood pressure and cigarette smoking:* Suppose that a tolerance of 2 years is specified and that the ages in the smokers group are 37, 38, 40, 45, and 50 years. (We shall assume that the smoker and nonsmoker groups are comparable on all other important variables.) A comparison reservoir twice the size ($r = 2$) of the smoking group consists of nonsmokers of ages 25, 27, 32, 36, 38, 40, 42, 43, 49, and 53 years. The estimated means of the two groups are 42.0 and 38.5 years, respectively. The ratio of the estimated variances, s_S^2/s_{NS}^2, is $0.37 = 29.50/79.78$.

The first step in forming the matches is to list the smokers and determine the corresponding comparison subjects who are within the 2-year tolerance from each smoker. For our example, this results in the possible pairing given in Table 6.1a.

Table 6.1a   Potential Caliper Matches for Example 6.1

Smokers	Nonsmokers
37	36, 38
38	38, 40
40	40, 42
45	43
50	49

It is clearly desirable to form matches for all the treatment subjects that are as close as possible. Thus the matched pairs shown in Table 6.1b would be formed.

Table 6.1b   Caliper-Matched Pairs

Smokers	Nonsmokers
37	36
38	38
40	40
45	43
50	49

Notice that if the 49-year-old nonsmoking subject had not been in the reservoir, we would not have been able to match all five smokers. We might then have decided to keep the first four matches and drop the 50-year-old smoker from the study. This results in a loss of precision, because the effective sample size is reduced. Alternatively, the tolerance could be increased to 3 years and the 50-year-old smoker matched with the 53-year-old nonsmoker. The latter approach does not result in lower precision, but the amount of bias may increase. Finally, had there been two or more comparison subjects with the same value of X, the match subject should be chosen randomly.

In Example 6.1 we knew the composition of the comparison reservoir before the start of the study. Often, however, this is not the case. Consider, for example, a study of the effect of specially trained nurses aids on patient recovery in a hospital. Such a study would require that the patients be matched on important confounding variables as they entered the hospital. When the comparison reservoir is unknown, the choice of a tolerance value that will result in a sufficient number of matched pairs can be difficult. The researcher cannot scan the reservoir, as we did in the example, and discover that the choice of ϵ is too small. For this reason, using caliper matching in a study where the comparison reservoir is unknown can result in matched sample sizes that are too small. In such a situation, the researcher can sometimes attempt to get a picture of the potential comparison population through records (i.e., historical data).

6.4.2   Appropriate Conditions

Caliper matching is appropriate regardless of the form of the relationship between the confounding and outcome variables (or risk variable in case-control studies). In this section we demonstrate how caliper matching is effective in reducing bias in both the linear and nonlinear cases.

Linear Case.   To understand how caliper matching works in the linear case, let us consider the estimate of the treatment effect or risk effect of smoking on blood pressure based on a 45-year-old smoker and a 43-year-old nonsmoker in Example 6.1. Assume that blood pressure is linearly related to age and that the relationships are the same for both groups, with the exception of the intercept values. Figure 6.5 represents this situation.

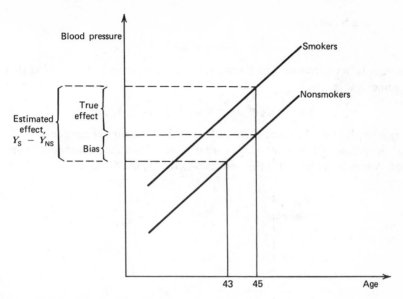

Figure 6.5 *Estimate of treatment effect—linear relationship.*

The estimate of the risk effect is shown by the large brace to the left in Figure 6.5. The amount of bias or distortion is shown by the small brace labeled "Bias." Relating this example to (6.3), we see that the bias is equal to the unknown regression coefficient β, multiplied by the difference in the values of the confounding variable. In the case of these two subjects, the bias is 2β. This is the maximum bias allowable under the specified tolerance for each individual estimate of the treatment effect, and consequently for the estimated treatment effect, based on the entire matched comparison group.

When we average the ages in Table 6.1b, we find that the mean age of the smokers is 42.0 years; of the nonsmokers comparison group, 41.2 years; and for the comparison reservoir, 38.5 years. Caliper matching thus reduced the difference in means from 3.5 ($= 42.0 - 38.5$) to 0.8 ($= 42.0 - 41.2$). In general, the extent to which the bias after matching, 0.8β in this case, is less than the maximum possible bias, 2β in this case, will depend on the quantities discussed in Section 6.2: the difference between the means of the two populations, the ratio of the variances and the size of the comparison reservoir as well as the tolerance.

Nonlinear Case. Let us now consider the case where the response and the confounding variable are related in a nonlinear fashion. To illustrate the effect of caliper matching in this situation, we shall assume that blood pressure is related to age squared. Algebraically this relationship between the response Y and age, X, can be written as

$$Y = \alpha + \beta X^2$$

where α is the intercept. The estimate of the treatment effect assuming that Y is numerical is

$$\overline{Y}_S - \overline{Y}_{NS} = \alpha_S - \alpha_{NS} + \beta \, (\overline{X_S^2} - \overline{X_{NS}^2}). \tag{6.5}$$

Hence any bias is a function of the difference in the means of age squared. Note that the means of the squared ages are different from the squares of the mean ages. Again let us visualize this relationship in Figure 6.6.

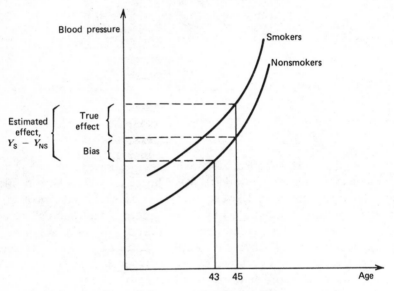

Figure 6.6 *Estimate of treatment effect—nonlinear relationship.*

The individual estimate of the treatment effect determined from the matched pair of a 45-year-old smoker and a 43-year-old nonsmoker is shown in Figure 6.6 by the large brace to the left. This estimate can be compared to the true treatment effect shown by the topmost smaller brace. The bias is the difference between the two and is indicated by the second small brace. From (6.5) we obtain the bias as

$$\beta(45^2 - 43^2) = \beta(176).$$

If we were using the matched groups from Example 6.1, upon averaging over the two groups we would find that the estimate of the treatment effect would be biased by the amount

$$\beta \, (\overline{X_S^2} - \overline{X_{NS}^2}) = \beta(69.6).$$

It is important to realize that equality of the means of the two groups is not enough to ensure an unbiased estimate of the treatment effect if the relationship between the response and the confounding variable is nonlinear. Equality of the means yields unbiased estimates only in the linear case.

6.4.3 Evaluation of Bias Reduction

So far we have only demonstrated how caliper matching can reduce the bias due to confounding. In this section we present theoretical results concerning the bias reduction one can expect using caliper matching in the linear case. The estimator of the treatment effect is the mean difference in response. The effectiveness of caliper matching and all other matching techniques is examined relative to estimating the treatment effect from random samples, where the confounding variable is not taken into account. (For a definition of the measure of effectiveness, the expected percent reduction in bias, see Cochran and Rubin, 1973.)

Table 6.2 gives an indication of the expected percent bias reduction for different tolerance values. The results are independent of the sample size and reservoir size. They were derived assuming that the initial difference between the two populations is less than 0.5 (i.e., $B_X < 0.5$), that the distributions of the confounding variable are normal, and that the outcome is linearly related to the confounding variable. Notice that the tolerance is specified in terms of a proportion, a, of a standard deviation. It appears that tight caliper matching (i.e., $a = 0.2$) can be expected to remove nearly all the bias in the treatment effect relative to random sampling. It also appears that the ratio of the variances (i.e., σ_1^2/σ_0^2) has a negligible effect on the percent reduction in bias.

Table 6.2   Percent Bias Reduction for Caliper Matching*

a	$\sigma_1^2/\sigma_0^2 = \frac{1}{2}$	$\sigma_1^2/\sigma_0^2 = 1$	$\sigma_1^2/\sigma_0^2 = 2$
0.2	99	99	98
0.4	96	95	93
0.6	91	89	86
0.8	86	82	77
1.0	79	74	69

Reprinted, by permission of the Statistical Publishing Society, from Cochran and Rubin (1973), Table 2.3.1.

* Tolerance $\epsilon = a\sqrt{(\sigma_1^2 + \sigma_0^2)/2}$.

One can use this table to get some indication of bias reduction to be expected for different tolerances if the values or estimates of the population variances are known and if $B_X < 0.5$. Suppose we knew that $\sigma_1^2/\sigma_0^2 = \frac{1}{2}$, where $\sigma_1^2 = 4$; then if we took $a = 0.8$, we could expect about 86% of the bias to be removed. The tolerance would be $0.8\sqrt{(4+8)/2} = 1.96$. If we used $a = 0.4$, we could expect to remove 96% of the bias over random sampling and the tolerance would be 0.98.

As we have mentioned previously, the major disadvantage of caliper matching is the need for the comparison reservoir to be large. In their theoretical work, Cochran and Rubin did not take into account the possibility that the desired number of matches would not be found from the comparison reservoir, although the probability of this occurrence is nonnegligible. Nor are the results known for distributions other than normal. Most likely, the results presented are applicable to symmetric distributions, but the case of skew distributions has not been investigated for caliper matching.

6.5   NEAREST AVAILABLE MATCHING

In some situations when caliper matching is performed with a small tolerance, there is a nonnegligible probability that some individuals cannot be matched. To avoid this problem, Rubin (1973a, b) developed a method known as nearest available pair matching. We shall refer to this matching procedure as *nearest available matching*. This method ensures that the desired number of matches are obtained by being less restrictive in deciding what a match is. A match is formed by finding the closest possible comparison subject for each individual in the treatment group from the yet-unmatched individuals in the comparison reservoir. Since nearest available matching does not use a fixed tolerance as does caliper matching, the reservoir does not have to be larger than the treatment group. However, the matches are not guaranteed to be as close as those found under caliper matching.

6.5.1   Methodology

There are three variants of nearest available matching, each based on a particular ordering of the subjects in the treatment group with respect to the confounding variable. The specification of the ordering completely defines the pair matching method. In one variant of the method, referred to as random-order nearest available matching, the N treatment subjects are randomly ordered on the values of the confounding variable, X. Let us denote these ordered values by X_{11} to X_{1N}. Starting with X_{11}, a match is defined as that subject from the comparison reservoir whose value X_{0j} is nearest X_{11}. The matches are therefore assigned to minimize $|X_{11} - X_{0j}|$ for all subjects in the comparison reservoir.

If there are ties (i.e., two or more comparison subjects for whom $|X_{11} - X_{0j}|$ is a minimum), the match is formed randomly. The nearest available partner for the next treatment subject with value X_{12} is then found from the remaining subjects in the reservoir. The matching procedure continues in this fashion until matches have been found for all N treatment subjects.

The other two variants of nearest available matching result from ranking the members of the treatment group on confounding variable values from the highest to the lowest (HL) value or from the lowest to the highest (LH) value. Matches are then sought starting with the first ranked treatment subject, as for the random-order version.

Example 6.2 Nearest available matching: Suppose that in a blood pressure study, there are three smokers with ages 40, 45, and 50, and five nonsmokers in the reservoir with ages 30, 32, 46, 49, and 55. In addition, suppose that the randomized order of the smokers' ages is 40, 50, and 45. Then the random-order nearest available matching technique will match the 40-year-old smoker with the 46-year-old nonsmoker, the 50-year-old smoker with the 49-year-old nonsmoker, and the 45-year-old smoker with the 55-year-old nonsmoker.

In the case of the other two variants, the following matches would be made: for HL, the 50-year-old with the 49-year-old, the 45-year-old with the 46-year-old, and the 40-year-old with the 32-year-old; for LH, the 40-year-old with the 46-year-old, the 45-year-old with the 49-year-old, and the 50-year-old with the 55-year-old. Notice in this example that each variant resulted in different matched pairs.

6.5.2 Appropriate Conditions

Nearest available matching is similar to caliper matching except that there is no fixed tolerance. Based on the prior discussion of caliper matching and some theoretical results, it follows that nearest available matching is effective in removing bias due to confounding if the relationship between the response and confounding variables is linear. For nonlinear relationships, no results are available.

The main difficulty in discussing what conditions are most appropriate for using nearest available matching is the fact that the reduction in bias is so strongly influenced by the closeness of the distributions of the treatment group and the comparison reservoir. If there is a large overlap between the two groups of subjects, nearest available matching will be very similar to caliper matching with a suitably large tolerance. If, however,there is a moderate to small amount of overlap, the desired number of matches will be found, but the final amount of bias in the estimate of the treatment effect may be large.

6.5.3 Evaluation of Bias Reduction

In selecting a particular nearest available matching procedure, an investigator may want to base his or her choice primarily on the percent reduction in bias obtainable. Assuming a linear relationship between the response and confounding variables, Cochran and Rubin (1973) performed a simulation study to determine which of the three nearest available matching estimators was least biased. They also assumed that the confounding variable was normally distributed with the mean of the treatment population greater than the mean in the comparison population ($\eta_1 > \eta_0$). Their results showed that the percent reduction in bias was largest for the low–high nearest available matching and smallest for the high–low variant.

Because nearest available matching does not guarantee as close matches as are possible with caliper matching, Rubin (1973a) also compared the closeness of the matches obtained by the three procedures as measured by the average of the squared error $(X_1 - X_0)^2$ within pairs. When the procedures were judged by this criterion, the order of performance was reversed. The HL nearest available matching had the lowest average squared error and the LH had the largest. This result is not too surprising, considering the relationship between the population means ($\eta_1 > \eta_0$). The HL procedure would start with the treatment subject who is likely to be the most difficult to match: namely, the one with the largest value of X. This would tend to minimize the squared within-pair difference.

Since the differences between the three matching procedures are small on both criteria, random-order nearest available matching appears to be a reasonable compromise. In Table 6.3, from Cochran and Rubin (1973), results of the percent reduction in bias are summarized for random-order nearest available matching as a function of the initial difference, the values of the ratio of the population variances, and sizes of the reservoir. Results for the number of matches $N = 25$ and $N = 100$ (not shown) differ only slightly from those for $N = 50$.

Table 6.3   Percent Bias Reduction for Random-Order Nearest Available Matching: X Normal; N = 50*

B_X / r	$\sigma_1^2/\sigma_0^2 = \frac{1}{2}$			$\sigma_1^2/\sigma_0^2 = 1$			$\sigma_1^2/\sigma_0^2 = 2$		
	$\frac{1}{4}$	$\frac{1}{2}$	1	$\frac{1}{4}$	$\frac{1}{2}$	1	$\frac{1}{4}$	$\frac{1}{2}$	1
2	99	98	84	92	87	69	66	59	51
3	100	99	97	96	95	84	79	75	63
4	100	100	99	98	97	89	86	81	71

Reprinted, by permission of the Statistical Publishing Society, from Cochran and Rubin (1973), Table 2.4.1.

* X = confounding variable; B_X = initial difference; r = ratio of the size of the comparison reservoir and the treatment group; σ_1^2 = variance of confounding variable in the treatment population; σ_0^2 = variance of confounding variable in the comparison population.

With this method, the percent reduction in bias decreases steadily as the initial difference between the normal distributions of the confounding variable increases from $\frac{1}{4}$ to 1. In contrast with results reported in Table 6.2 for caliper matching, the percent reduction in bias does depend on the ratio of the population variances. Based on Table 6.3, random-order nearest available matching does best when $\sigma_1^2/\sigma_0^2 = \frac{1}{2}$. When $\eta_1 > \eta_0$ and $\sigma_0^2 > \sigma_1^2$, large values of the confounding variable in the treatment group, the ones most likely to cause bias, will receive closer partners out of the comparison reservoir than if $\sigma_0^2 < \sigma_1^2$.

Investigators planning to use random-order nearest available matching can use Table 6.3 to obtain an estimate of the expected percent bias reduction. Suppose an estimate of the initial difference B_X is $\frac{1}{2}$, with $\sigma_1^2/\sigma_0^2 = 1$, and it is known that the reservoir size is 3 times larger than the treatment group ($r = 3$). It follows that random-order nearest available matching results in an expected 95% reduction in bias.

6.6   STRATIFIED MATCHING

Stratified matching is an appropriate pair matching procedure for categorical confounding variables. If, like sex or religious preference, the variable is truly categorical, with no underlying numerical distribution, the matches are exact and no bias will result. Often, however, the confounding variable is numerical but the investigator may choose to work with the variable in its categorical form. Suppose, for example, that in the study of smoking and blood pressure, all the subjects were employed and that job anxiety is an important confounding variable. The investigator has measured job anxiety by a set of 20 true–false questions so that each subject can have a score from 0 to 20. Such a factor is very difficult to measure, however, and the investigator may decide that it is more realistic and more easily interpretable to simply stratify the range of scores into low anxiety, moderate anxiety, and high anxiety. Having formed these three strata, the investigator can now randomly form individual pair matches within each stratum. An example of this procedure in the case of multiple confounding variables is given in Section 6.11.

The only theoretical paper discussing the bias reduction properties of stratified matching is that of McKinlay (1975). She compared stratified matching to various stratification estimators (Section 7.6) for a numerical confounding variable converted to a categorical variable. She considered various numbers of categories and a dichotomous outcome. She found that the estimator of the odds ratio from stratified matched samples had a larger mean squared error and, in some of the cases considered, a larger bias than did the crude estimator, which ignores the confounding variable. (Stratified matching is compared with stratification in Section 13.2.2.) The mean squared error results are due in part to the loss of precision caused by an inability to find matches for all the treatment subjects. This point is considered further in Section 13.2.

6.7 FREQUENCY MATCHING

Frequency matching involves stratifying the distribution of the confounding variable in the treatment group and then finding comparison subjects so that the number of treatment and comparison subjects is the same within each stratum. This is not a pair matching method, and the number of subjects may differ across strata.

For the sake of illustration we shall concentrate on the case of a numerical response. This will allow us to demonstrate more easily how frequency matching helps to reduce the bias. Because frequency matching is equivalent to stratification with equal numbers of comparison and treatment subjects within each stratum, we leave the discussion of the various choices of estimators in the case of a dichotomous response to Chapter 7.

6.7.1 Methodology

Frequency matching is most useful when one does not want to deal with pair matching on a numerical confounding variable or an ordinal measure of an underlying numerical confounding variable. An example of the latter situation is initial health care status, where the categories reflect an underlying continuum of possible statuses. In either case, the underlying distribution must be stratified. Samples are then drawn either randomly or by stratified sampling from the comparison reservoir in such a way that there is an equal number of treatment and comparison subjects within each stratum. Criteria for choosing the strata are discussed in Section 6.7.3 after we have presented the estimator of the treatment effect.

Example 6.3 Frequency matching: Let us consider the use of frequency matching in the smoking and blood pressure study. Suppose that the age distribution of the smokers was stratified into 10-year intervals as shown on the first line of Table 6.4, and that 100 smokers were distributed across the strata as shown on the second line of the table. The third line of the table represents the results of a random sample of 100 nonsmokers from the comparison reservoir. Notice that since frequency matching requires the sample sizes to be equal within each stratum, the investigator needs to draw more nonsmokers in all strata except for ages 51 to 60 and 71 to 80. In these two strata the additional number of nonsmokers would be dropped from the study on a random basis. (Note that stratified sampling, if possible, would have avoided the problem of too few or too many persons in a stratum.)

Table 6.4   Smokers and Nonsmokers Stratified by Age

Age	11–20	21–30	31–40	41–50	51–60	61–70	71–80	Total
Smokers	1	3	10	21	30	25	10	100
Nonsmokers	0	2	8	20	32	20	18	100

6.7.2   Appropriate Conditions

Frequency matching is relatively effective in reducing bias in the parallel linear response situation provided that enough strata are used. We shall explain this by means of simple formulas for the estimator of the treatment effect assuming a numerical response.

Recall from Section 6.1 that we can represent the linear relationship between the response Y and the confounding variable X by

$$Y_1 = \alpha_1 + \beta X_1 \qquad \text{in the treatment group} \qquad (6.6)$$

$$Y_0 = \alpha_0 + \beta X_0 \qquad \text{in the comparison group.}$$

In general, the estimator of the treatment effect in the kth stratum is

$$\overline{Y}_{1k} - \overline{Y}_{0k} = (\alpha_1 - \alpha_0) + \beta(\overline{X}_{1k} - \overline{X}_{0k}), \qquad (6.7)$$

where a bar above the variables indicates the mean calculated for the kth stratum. The bias in the kth stratum is $\beta(\overline{X}_{1k} - \overline{X}_{0k})$.

Clearly, the maximum amount of distortion in the estimate from the kth stratum occurs when $\overline{X}_{1k} - \overline{X}_{0k}$ is maximized. The maximum value is then β times the width of the kth stratum.

One overall estimate of the treatment effect is the weighted combination of the individual strata differences in the response means:

$$\overline{Y}_1 - \overline{Y}_0 = \frac{1}{N} \sum_{k=1}^{K} n_k(\overline{Y}_{1k} - \overline{Y}_{0k}), \qquad (6.8)$$

where n_k is the number of treatment or comparison subjects in the kth stratum ($k = 1, 2, \ldots, K$) and N is the total number of treatment subjects. Rewriting (6.8) in terms of treatment effect and regression coefficients, we obtain, using (6.7),

$$\overline{Y}_1 - \overline{Y}_0 = \frac{1}{N} \sum_{k=1}^{K} n_k[\alpha_1 - \alpha_0 + \beta(\overline{X}_{1k} - \overline{X}_{0k})]$$

$$= (\alpha_1 - \alpha_0) + \frac{1}{N} \sum_{k=1}^{K} n_k \beta(\overline{X}_{1k} - \overline{X}_{0k}). \qquad (6.9)$$

From (6.9) we see that the amount of bias reduction possible using frequency matching is determined by the difference in the distributions of the two groups within each stratum. This, in turn, is a function of the manner in which the strata were determined. The more similar the distributions of the treatment and comparison populations are within each stratum, the less biased the individual estimates of the treatment effect will be.

6.7.3 Evaluation of Bias Reduction

Assuming that both distributions of the confounding variable are normal with equal variances but the mean of the treatment population is zero and the mean of the comparison population is small but nonzero, Cox (1957) derived the percent reduction in bias for strata with equal number of subjects. Cochran and Rubin (1973) extended Table 1 of Cox, and these results are given in Table 6.5. The strata are based on the distribution of the treatment group.

Table 6.5   Percent Bias Reduction with Equal-Sized Strata in Treatment Population: X Normal

Number of strata:	2	3	4	5	6	8	10
% reduction in bias:	64	79	86	90	92	94	96

Reprinted, by permission of the Statistical Publishing Society, from Cochran and Rubin (1973), Table 4.2.1.

These percentages are at most 2% lower than the maximum amount of bias reduction possible using strata with an unequal number of subjects. Cochran (1968) extended these calculations for some nonnormal distributions: the chi-square, the t, and the beta, and he concluded that the results given in Table 6.5 can be used as a guide to the best boundary choices even when the confounding variable is not normally distributed.

From this information we can conclude that if the distributions of the confounding variable are approximately normal and differ only slightly in terms of the mean, and based on the distribution of the treatment group we form four strata with equal numbers of subjects, we can expect to reduce the amount of bias in the estimate of the treatment effect by 86%.

We stated at the beginning of Section 6.7.2 that frequency matching was relatively effective in reducing the bias in the linear parallel situation. No theoretical work has been done for the nonlinear parallel situation. Frequency matching does, however, have the advantage of allowing one to use the analysis of variance to test for interactions. One can test for parallelism as well as linearity, thus determining whether frequency matching was appropriate.

6.8   MEAN MATCHING

A simple way of attempting to equate the distributions of the confounding variable in the study samples is to equate their means. This is called *mean matching* or *balancing*. The members of the comparison group are selected so that $|\overline{X}_1 - \overline{X}_0|$ is as small as possible. Although mean matching is very simple to employ, it depends strongly on the assumption of a linear parallel response relationship and we therefore do not recommend its use. One can employ analysis of covariance (Chapter 8) in this case and achieve greater efficiency. We include the following discussion of mean matching so that the reader can understand the basis for our recommendation.

6.8.1   Methodology

There is more than one way to form matches in mean matching. However, the only algorithm which is guaranteed to find the comparison group that minimizes $|\overline{X}_1 - \overline{X}_0|$ is to calculate $\overline{X}_0$ for all possible groups of size N from the comparison reservoir. This is generally far too time-consuming. An easier algorithm uses partial means, and we shall demonstrate its use with the following example.

Example 6.4   Mean matching: Suppose that we decided to use mean matching on age in the blood pressure study, where we have three smokers, aged 40, 45, and 50 years. First, we would calculate the mean age of the smokers, which is 45 years ($\overline{X}_S = 45$). Next, we would select successive subjects from the nonsmokers such that the means of the nonsmokers ages, calculated after the selection of each subject (partial means), are as close as possible to 45. Suppose that the nonsmokers in the comparison reservoir have the following ages: 32, 35, 40, 41, 45, 47, and 55 years. The first nonsmoker selected as a match would be age 45; the second subject selected would be 47 years old, since the partial mean, $(45 + 47)/2 = 46$, is closest to 45. The last nonsmoker to be selected would be 41 years of age, again since the partial mean, $(2/3)(46) + (1/3)(41) = 44.3$, is closest to $\overline{X}_S$. Note that this algorithm did not minimize $|\overline{X}_S - \overline{X}_{NS}|$, since choosing the nonsmokers aged 35, 45, and 55 would give equality of the two sample mean ages $[(35 + 45 + 55)/3 = 45]$.

6.8.2   Appropriate Conditions

Mean matching can be very effective in reducing bias in the case of a parallel linear response relationship. Suppose in the blood pressure example that the population means η_S and η_{NS} for smokers and nonsmokers were 50 and 45, respectively. Then, for large enough random samples, we might expect to find that $\overline{X}_{NS} = 45$ and $\overline{X}_S = 50$.

From (6.3) it follows that the estimated treatment effect is biased by an amount equal to $\beta\,(\overline{X}_S - \overline{X}_{NS}) = 5\beta$. However, if mean matching had been used to reduce $|\overline{X}_S - \overline{X}_{NS}|$ to, say, 0.7, as in Example 6.4, then the bias in $(\overline{Y}_S - \overline{Y}_{NS})$ would have been reduced by 86% (= 4.3/5.0). (The initial difference in the means due to random sampling is 5.0.)

Mean matching is *not* effective in removing bias in the case of a parallel nonlinear response relationship (see Figure 6.7). Assume that in another blood pressure study three smokers of ages 30, 35, and 40 years were mean-matched with three nonsmokers of ages 34, 35, and 36 years, respectively. Their blood pressures are denoted by × in Figure 6.7. Notice that unlike the previous linear situations, $\overline{Y}_S$ and $\overline{Y}_{NS}$ do not correspond to the mean ages $\overline{X}_S$ and $\overline{X}_{NS}$. They will both be greater than the values of Y which correspond to the means due to the nonlinearity. Here $(\overline{Y}_S - \overline{Y}_{NS})$ is an overestimate of the treatment effect. The estimate should be equal to the length of the vertical line, which represents the treatment effect. In general, the greater the nonlinearity, the greater the overestimation or bias will be, in general.

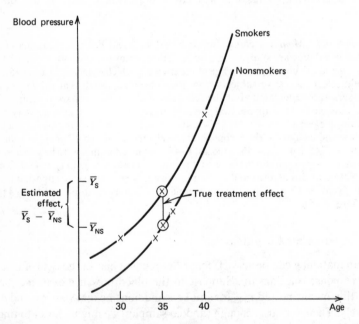

Figure 6.7 *Mean matching in a nonlinear parallel relationship.*
×, *blood pressure for a specific age;* ⊗, *blood pressure corresponding to mean age in either group.*

6.8.3   Evaluation of Bias Reduction

Cochran and Rubin (1973) have investigated the percentage of bias reduction possible using the partial mean algorithm presented in Section 6.8.1 under the assumptions of a linear parallel relationship, a normally distributed confounding variable, and a sample size of 50 in the treatment group. They found that, except in the cases where the initial difference $B_X = 1$, mean matching removes essentially all the bias. In addition, its effectiveness increases with the size of the comparison reservoir. The bias that results from improper use of mean matching (i.e., in nonlinear cases) has not been quantified.

6.9   ESTIMATION AND TESTS OF SIGNIFICANCE

In this section we indicate the appropriate tests of significance and estimators of the treatment effect for each matching technique. Because the choice of test and estimator depends on the form of the outcome variable, we begin with the numerical case followed by the dichotomous case. Also, in keeping with the general intent of this book, we do not give many details on the test statistics but rather cite references in which further discussion may be found. The tests and estimators for frequency-matched samples are the same as for stratification and are discussed in greater detail in Chapter 7.

In the case of a numerical outcome variable for which one of the pair matching methods (caliper, nearest available, or stratified) has been used, the correct test of significance for the null hypothesis of no treatment effect is the paired-t test (see Snedecor and Cochran, 1967, Chap. 4). This test statistic is the ratio of the mean difference, which is the estimate of the treatment effect, to its standard error. The difference between the paired-t test and the usual t test for independent (nonpaired) samples is in the calculation of the standard error.

If in the case of a numerical outcome variable, frequency matching has been used, the standard t test is appropriate, with the standard error determined by an analysis of variance. (See Snedecor and Cochran, 1967, Chap. 10, for a discussion of the analysis of variance.) The treatment effect is estimated by the mean difference. If, however, the within-stratum variances are not thought to be equal, then, as in the case of stratification, one should weight inversely to the variance (see Section 7.7 and Kalton, 1968). In the case of mean matching, the correct test is again the t test. The standard error, however, must be calculated from an analysis of covariance (see Greenberg, 1953).

When the outcome variable is dichotomous, as discussed in Chapter 3, the treatment effect may be measured by the difference in proportions, the relative risk, or the odds ratio. The estimator of the difference in rates is the difference between the sample proportions, $p_1 - p_0$. This is an unbiased estimator if the matching is exact. For estimating the odds ratio, the stratification estimators appropriate for large numbers of strata are applicable (see Section 7.6.1), with each pair comprising a stratum. In this case the conditional maximum likelihood estimator is easy to calculate and is identical to the Mantel–Haenszel (1959) estimator. For each pair (stratum), a 2×2 table can be created. For the jth pair, we have four possible outcomes:

		Control Subject	
		1	0
Treatment Subject	1	a_j	b_j
	0	c_j	d_j

For example, $b_j = 1$ if, in the jth pair, the outcome for the control subject is 0 and for the treatment subject it is 1. The estimator of the odds ratio, ψ, is then $\hat{\psi} = \Sigma_j b_j / \Sigma_j c_j$. The estimator will be approximately unbiased if the matching is exact and the number of pairs is large.

Because of the relationship between these measures of the treatment effect (difference of proportions, relative risk, and odds ratio) under the null hypothesis of no treatment effect (Section 3.1), McNemar's test can be used in the case of pair-matched samples, regardless of the estimator (see Fleiss, 1973, Chap. 8). Similarly, when frequency matching is used, we have a choice of tests, such as Mantel–Haenszel's or Cochran's test, regardless of the estimator (see Fleiss, 1973, Chap. 10). Since the analysis of a frequency-matched sample is the same as an analysis by stratification, the reader is referred to Chapter 7 for a more detailed discussion.

6.10 MULTIVARIATE MATCHING

So far we have limited the discussion of matching to a single confounding variable. More commonly, however, one must control simultaneously for many confounding variables. To date, all research has been on multivariate pair matching methods. To be useful, a multivariate matching procedure should create close individual matches on all variables. In addition, ideally, as in the univariate case, the procedure should not result in the loss of many subjects because of a lack of suitable matches. The advantage of constructing close individual matches, as in the univariate case, is that with perfectly matched pairs the matching variables are perfectly controlled irrespective of the underlying model relating the outcome to the risk and confounding variables.

Discussions of multivariate matching methods in the literature are quite limited. References include Althauser and Rubin (1970), for a discussion of an applied problem; Cochran and Rubin (1973), for a more theoretical framework; Rubin (1976a, b), for a discussion of certain matching methods that are equal percent bias reducing (EPBR); Carpenter (1977), for a discussion of a modification of the Althauser–Rubin approach; and Rubin (1979), for a Monte Carlo study comparing serveral multivariate methods used alone or in combination with regression adjustment.

In the following sections we first discuss straightforward generalizations of univariate caliper and stratified matching methods to the case of multiple confounding variables. The methods included are multivariate caliper matching, and multivariate stratified matching. Then we discuss metric matching methods wherein the objective is to minimize the distance between the confounding variable measurements in the comparison and treatment samples. Several alternative distance definitions will be presented.

Next we discuss discriminant matching. This matching method reduces the multiple confounding variables to a single confounding variable by means of the linear discriminant function. Any univariate matching procedure can then be applied to the linear discriminant function.

In trying to rank the multivariate matching techniques according to their ability to reduce the bias, one is faced with the problem of how to combine the reduction in bias due to each confounding variable into a single measure so that the various methods can be compared. For example, the effectiveness of caliper matching depends, in part, on the magnitudes of all the tolerances that must be chosen.

To partially circumvent this problem of constructing a single measure of bias reduction, Rubin (1976a, b; 1979) introduced the notion of matching methods of the equal percent bias reducing (EPBR) type. For the linear case, Rubin showed that the percent bias reduction of a multivariate matching technique is related to the reduction in the differences of the means of each confounding variable, and that if the percent reduction is the same for each variable, that percentage is the percent reduction for the matching method as a whole. EPBR matching methods are techniques used to obtain equal percent reduction on each variable and, hence, guarantee a reduction in bias.

Discriminant matching and certain types of metric matching have the EPBR property, so that we can indicate which of these EPBR methods can be expected to perform best in reducing the treatment bias in the case of a linear response surface.

6.10.1 Multivariate Caliper Matching

Multivariate caliper matching, like its univariate counterpart, is effective in reducing bias provided that the tolerances used for each confounding variable are small and the comparison reservoir is large, generally much larger than in the univariate case.

Suppose that there are L confounding variables. A comparison subject is considered to be a match for a treatment subject when the difference between their measured lth confounding variable ($l = 1, 2, \ldots, L$) is less than some specified tolerance, ϵ_l (i.e., $|X_{1l} - X_{0l}| \leq \epsilon_l$) for all l.

Example 6.5   Multivariate caliper matching: Consider a hypothetical study comparing two therapies effective in reducing blood pressure, where the investigators want to match on three variables: previously measured diastolic blood pressure, age, and sex. Such confounding variables can be divided into two types: categorical variables, such as sex, for which the investigators may insist on a perfect match ($\epsilon = 0$); and numerical variables, such as age and blood pressure, which require a specific value of the caliper tolerances. Let the blood pressure tolerance be specified as 5 mm Hg and the age tolerance as 5 years. Table 6.6 contains measurements of these three confounding variables. (The subjects are grouped by sex to make it easier to follow the example.)

Table 6.6   Hypothetical Measurements on Confounding Variables for Example 6.6

Treatment Group				Comparison Reservoir			
Subject Number	Diastolic Blood Pressure (mm Hg)	Age	Sex	Subject Number	Diastolic Blood Pressure (mm Hg)	Age	Sex
1	94	39	F	1	80	35	F
2	108	56	F	2	120	37	F
3	100	50	F	3	85	50	F
4	92	42	F	4	90	41	F
5	65	45	M	5	90	47	F
6	90	37	M	6	90	56	F
				7	108	53	F
				8	94	46	F
				9	78	32	F
				10	105	50	F
				11	88	43	F
				12	100	42	M
				13	110	56	M
				14	100	46	M
				15	100	54	M
				16	110	48	M
				17	85	60	M
				18	90	35	M
				19	70	50	M
				20	90	49	M

In this example there are 6 subjects in the treatment group and 20 subjects in the comparison reservoir. Given the specified caliper tolerances, the first subject in the treatment group is matched with the fourth subject in the comparison reservoir. The difference between their blood pressures is 4 units, their ages differ by 2 years, and both are females. We match the second treatment subject with the seventh comparison subject since their blood pressures and sex agree exactly and their ages differ by only 3 years. The remaining four treatment subjects, subjects 3, 4, 5, and 6, would be matched with comparison subjects 10, 8, 19, and 18, respectively. Notice that if the nineteenth comparison subject were not in the reservoir, the investigator would have to either relax the tolerance on blood pressure, say to 10 mm Hg, or discard the fifth treatment subject from the study.

Expected Bias Reduction.   Table 6.2 gives the expected percent of bias reduction for different tolerances assuming a single, normally distributed confounding variable and a linear and parallel response relationship. Table 6.2 can also be used in the case of multiple confounding variables if these variables or some transformation of them are normally and independently distributed, and if the relationship between the outcome and confounding variables is linear and parallel. The expected percent of bias reduction is then a weighted average of the percent associated with each variable.

If the investigators know (a) the form of the linear relationship, (b) the population parameters of the distribution of each of the confounding variables, and (c) that the confounding variables or some transformation of them are independent and normally distributed, then the best set of tolerances in terms of largest expected treatment bias reduction in Y could theoretically be determined by evaluating equation (5.1.5) in Cochran and Rubin (1973) for several combinations of tolerances. In practice, this would be very difficult to do.

6.10.2   Multivariate Stratified Matching

The extension of univariate stratified matching to the case of multiple confounding variables is straightforward. Subclasses are formed for each confounding variable, and each member of the treatment group is matched with a comparison subject whose values lie in the same subclass on all confounding variables.

Example 6.6   Multivariate stratified matching: Consider again the blood pressure data presented in Table 6.6. Suppose that the numerical confounding variable, diastolic blood pressure, is categorized as ≤ 80, 81–94, 95–104, and ≥ 105, and age as 30–40, 41–50, and 51–60. Including the dichotomous variable, sex, there are in total ($4 \times 3 \times 2 =$) 24 possible subclasses into which a subject may be classified. In Table 6.7 we enumerate the 12 possible subclasses for males and females separately. Within each cell we have listed the subject numbers and indicated by the subscript t those belonging to the treatment group.

**Table 6.7   Stratification of Subjects on Confounding Variables
in Example 6.6[a]**

Diastolic Blood Pressure	Age		
	30–40	41–50	51–60
		Males	
−80		5_t, 19	
81–94	6_t, 18	20	17
95–104		12, 14	15
105-		16	13
		Females	
−80	1, 9		
81–94	1_t	4_t, 3, 4, 5, 8, 11	6
95–104		3_t	
105-	2	10	2_t, 7

[a] Within each cell the subject number from Table 6.6 is given. Those with a subscript t are the treatment group subjects.

With this stratification, the second treatment subject is matched with the seventh comparison subject. The fifth treatment subject would be matched with the nineteenth comparison subject and the fourth treatment subject would be randomly matched with one of comparison subjects 3, 4, 5, 8, or 11. The last treatment subject would be matched with the eighteenth comparison subject. Subjects 1 and 3 in the treatment group do not have any matches in the comparison reservoir and must therefore be omitted from the study, or else the subclass boundaries must be modified.

It should be clear from this simple example that as the number of confounding variables increases, so does the number of possible subclasses, and hence the larger the comparison reservoir must be in order to find an adequate number of matches.

The expected number of matches for a given number of subclasses and given reservoir size r have been examined by McKinlay (1974) and Table 6.8 presents a summary of her results. The number of categories in Table 6.8 equals the product of the number of subclasses for each of the L confounding variables. In McKinlay's terminology we had 24 categories in Example 6.6. Her results are based on equal as well as markedly different joint distributions of the L confounding variables in the treatment and comparison populations (see McKinlay, 1974, Table 1, for the specific distributions). For example, in a study with 20 subjects in the treatment group and 20 in the comparison reservoir, stratified matching on 10 categories where the confounding variable distributions in the two populations are exactly the same will result in about 66 percent of the treatment group being matched (i.e., only 13 suitable comparison subjects would be expected to be found). Clearly, large reservoirs are required if multivariate

Table 6.8   Expected Percentages of Matches in Multivariate Stratified Matching

N, Size of Treatment Group	r	Same Distribution 10 Categories	Same Distribution 20 Categories	Different Distribution 10 Categories	Different Distribution 20 Categories
20	1	66.0	53.0	55.0	43.5
	2.5	94.0	84.5	84.5	72.5
	5	98.5	96.0	96.5	89.0
	10	100.0	99.0	99.5	96.5
50	1	78.0	68.6	62.4	55.2
	2	97.0	91.6	86.6	78.0
	4	99.8	98.8	98.0	92.8
	10	100.0	100.0	100.0	99.0
100	1	84.3	77.3	65.3	60.5
	2	99.1	96.8	90.3	83.7
	5	100.0	99.9	99.8	97.2

Adapted, by permission of the Royal Statistical Society, from McKinlay (1974), Tables 2 and 3.

stratified matching is to be used effectively. With 20 treatment subjects one would need more than 100 comparison subjects for matching with only negligible loss of treatment subjects.

No information is available on the bias reduction one can expect for a given reservoir size, r, and given population parameters of the joint distribution of the L confounding variables in the treatment and comparison populations.

6.10.3   Minimum Distance Matching.

Both multivariate caliper matching and stratified matching are straightforward extensions of univariate techniques in that a matching restriction exists for each variable. In this section we discuss *minimum distance matching* techniques that take all of the confounding variables into account at one time, thus reducing multiple matching restrictions to one. For two subjects to be a match, their confounding variable values must be close as defined by same distance measure. The matching can be done with a "fixed" tolerance, as in univariate caliper matching, or as nearest available matching. We begin with the fixed tolerance case. Because distance is defined by a distance function or metric, these techniques are also referred to as *metric matching*.

One distance function is Euclidean distance which is defined as

$$\sum_{l=1}^{L} (X_{1l} - X_{0l})^2, \qquad (6.10)$$

where X_{il} is the value of the lth confounding variable for a subject in the treatment ($i = 1$) or the comparison ($i = 0$) group. A major problem with the use of Euclidean distance is that the measure (6.10) and hence choice of matched subjects strongly depend on the scale used for measuring the confounding variables. For example, measuring a variable in centimeters rather than in meters would increase that variable's contribution to the Euclidean distance 10,000-fold.

A common technique for eliminating this problem of choice of scale is to convert all variables to *standardized scores*. A standardized score (Z) is the observed value of a confounding variable (X), divided by that confounding variable's standard deviation (s): $Z = X/s$. Equation (6.10) then would become

$$\sum_{l=1}^{L} (Z_{1l} - Z_{0l})^2, \qquad (6.11)$$

where Z_{il} is the standardized score of the lth confounding variable ($l = 1, \ldots, L$) for a subject in the treatment ($i = 1$) or the comparison ($i = 0$) group. Use of (6.11) as a matching criterion has been termed *circular matching* (Carpenter, 1977).

To better understand circular matching and its relation to multivariate caliper matching, consider the case of two confounding variables shown in Figure 6.8. Suppose that the two confounding variables have been transformed to standardized scores. Point A is a treatment subject with standardized scores of a_1 for the first confounding variable and a_2 for the second. If we were to use multivariate caliper matching with a common tolerance ϵ, we would search for a comparison subject with a standardized score of the first confounding variable in the interval $[a_1 - \epsilon, a_1 + \epsilon]$, and at the same time, a value of the second standardized score in the interval $[a_2 - \epsilon, a_2 + \epsilon]$. Thus the search is for a comparison subject like subject B, with confounding variable values within the *square* shown in Figure 6.8.

In circular matching with tolerance ϵ, the search is for a comparison subject whose confounding variable values satisfy

$$(Z_{11} - Z_{01})^2 + (Z_{12} - Z_{02})^2 \leq \epsilon,$$

that is, for values in the *circle* of radius ϵ centered at A. In Figure 6.8, subject B would not be a match for subject A if circular matching with tolerance ϵ were used.

There have been several suggestions for calculating the standard deviation to be used in the standardized scores. Cochran and Rubin (1973) suggest using

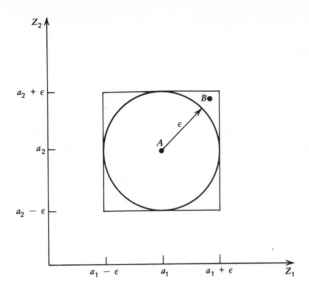

Figure 6.8 *Caliper matching on standardized scores.* ϵ = *tolerance.*

the standard deviations calculated from the comparison group only, while Smith et al. (1977) suggest using only the treatment group standard deviations. The advantage of the latter suggestion is that the standard deviations may be calculated before identifying the comparison subjects. Finally, a pooled estimate of the standard deviation can be used if one believes that the variances of the two groups are similar. All three suggestions suffer from the restriction that the measurements for calculating the standard deviations must be available prior to any matching.

Equation (6.11) can be rewritten as

$$\sum_{l=1}^{L} (X_{1l} - X_{0l})^2/s_l^2.$$

As can be seen from this, circular matching only takes the variances of the confounding variables into account and neglects possible correlations between these variables. An alternative metric matching technique which takes correlations into account is *Mahalanobis metric matching,* based on the following measure of distance in matrix notation:

$$(X_1 - X_0)'S^{-1}(X_1 - X_0), \tag{6.12}$$

where S is the matrix of sample variances and covariances (specifically, the pooled within-sample covariance matrix) and X represents a column vector of values of the confounding variables. This distance function can be used in the same way as Euclidean distance using a fixed tolerance to define a match.

Circular and Mahalanobis metric matching with tolerance ϵ can result in a loss of information, however, since there is no guarantee of matching all treatment subjects, even when the comparison reservoir is large. One approach to overcome this potential loss of treatment subjects is to generalize the method of nearest available matching (Section 6.5). Cochran and Rubin (1973) suggest randomly ordering the treatment subjects and then assigning as a match the comparison subject who is not yet matched and who is nearest as measured by some distance function, such as (6.10), (6.11), or (6.12). Such methods are called *nearest available metric matching* methods. Smith et al. (1977) proposed nearest available circular matching. Rubin (1979) compared the percent bias reduction of nearest available Mahalanobis metric matching with that of nearest available discriminant matching (Section 6.10.4). Rubin's study is discussed in Section 6.10.5.

6.10.4 Discriminant Matching

Another approach for dealing with multiple confounding variables is to transform the many variables to a single new variable and then to apply a univariate matching procedure to this single variable. One such transformation is the *linear discriminant function*. Basically, the linear discriminant function is a linear combination of the confounding variables that best predicts group membership. In a sense, it is the variable on which the groups differ the most.* By matching on this single variable, it is hoped to achieve the maximum amount of bias reduction. Using one of the univariate matching procedures described above on the linear discriminant function, those cases will be selected whose discriminant function values are the closest. For more detailed references on discriminant matching, see Cochran and Rubin (1973) and Rubin (1976a, b; 1979). Snedecor and Cochran (1967, Chap. 13) show how to use multiple regression to calculate the discriminant function.

6.10.5 Multivariate Matching with Linear Adjustment

Rubin (1979) empirically examined the nearest available Mahalanobis metric and nearest available discriminant matching methods, alone and in combination with regression adjustment on the matched pair differences for various sampling situations, and for various underlying models, both linear and nonlinear. (See Section 13.3.1 for a discussion of matching with regression adjustment.) Rubin selected these two methods because, under certain distributional assumptions,

---

* A good survey paper on discriminant analysis is that of Lachenbruch (1979). In this paper a discussion is given of discriminant analysis on numerical variables, categorical variables, and multivariate data containing both numerical and categorical variables.

they are equal percent bias reducing (EPBR); that is, they yield the same percent reduction in bias for each matching variable. As a result, this percent bias reduction is a straightforward criterion of how well the EPBR matching method has reduced bias in the estimate of the treatment effect.

The broad conclusion of Rubin (1979) is that nearest available pair matching using the Mahalanobis metric, together with regression adjustment on the matched pair differences, is an effective plan for controlling the bias due to the confounding variables, even for moderately nonlinear relationships.* Over a wide range of distributional conditions used in his Monte Carlo study, this metric matching method reduced the expected squared bias by an average of 12% more than did random sampling with no matching. (Notice that for univariate matching methods, the results given in Tables 6.2, 6.3, and 6.5 all relate to percent bias reduction and not to percent squared bias reduction.) This metric matching reduces more than 90% of the squared bias. Without regression adjustment, nearest available discriminant matching is equivalent to nearest available Mahalanobis metric matching, although Rubin finds that Mahalanobis metric matching is more robust against alternative model and distributional specifications.

6.11 MULTIPLE COMPARISON SUBJECTS

Occasionally, matched samples may be generated by matching each treatment subject with more than one comparison subject. Matching with multiple controls is especially advantageous when the number of potential comparison subjects is large relative to the number of available treatment subjects or when the unit cost for obtaining the comparison subjects is substantially lower than that of obtaining treatment subjects.

The present discussion concentrates on the dichotomous outcome case. Assume that each treated subject is matched with the same number, say q, of comparison subjects. The selection of a particular multivariate matching procedure should be based on the same principles explained in previous sections for a single comparison subject. For an example of pair matching using multiple controls, see Haddon et al. (1961).

Let the data from the jth matched group, $j = 1, 2, \ldots, N$, be represented in terms of a 2×2 frequency table, Table 6.9. Here $a_j = 1$ if the treatment subjects have the outcome factor present and $a_j = 0$ otherwise; b_j is the number of comparison subjects who have the outcome factor present.

---

* For univariate matching, an extreme example of a moderately nonlinear relationship is $Y = \exp(X)$, whereas $Y = \exp(X/2)$ is more reasonable. In multivariate matching, a similar statement can be made.

Table 6.9   Multiple Comparison Subjects Data from jth Matched Sample

Outcome	Treatment Subjects	Comparison Subjects	Total
Factor present (= 1)	a_j	b_j	$a_j + b_j$
Factor absent (= 0)	$1 - a_j$	$q - b_j$	$1 + q - (a_j + b_j)$
Total	1	q	$1 + q$

For simplicity, let us make the following definitions:

$$A = \sum_{j=1}^{N} a_j,$$

where A is the total number of treatment subjects who have the outcome factor present, and

$$B = \sum_{j=1}^{N} b_j,$$

where B is the total number of control subjects who have the outcome factor present. Therefore, the rate at which the outcome factor is present among the treatment group is $p_1 = A/N$, and the rate at which it is present among the comparison group is $p_0 = B/qN$. The difference in rates, as a measure of treatment effect, is then estimated by $p_1 - p_0$.

To estimate the odds ratio, each set of $q + 1$ subjects is considered a stratum, and estimators appropriate to stratified samples are applied. (See Section 7.6.1 for a more detailed discussion of estimators of the odds ratio that are appropriate when the number of strata becomes large.) Two such estimators, the conditional maximum likelihood and Mantel–Haenszel estimators, are given by Miettinen (1970). For $q \geq 3$, the conditional maximum likelihood estimator becomes difficult to use because it requires an iterative solution. For the case of exact matching, the conditional maximum likelihood estimator will be approximately unbiased for large N; Miettinen conjectures that the same is true for the Mantel–Haenszel estimator. No comparison of these two estimators as applied to multiple comparison subjects has been made. McKinlay's (1978) results for stratification (Section 7.6.2) imply that the Mantel–Haenszel estimator will be less biased than the conditional maximum likelihood estimator will be. For the case of a single comparison subject for each treatment subject ($q = 1$), the two estimators are identical (and are given in Section 6.9).

To test the null hypothesis of no treatment effect, we wish to consider the difference between p_1 and p_0. An appropriate test statistic is

$$T = \frac{p_1 - p_0}{SE(p_1 - p_0)}$$

$$= \frac{qA - B}{(q + 1)(A + B) - \sum_{j=1}^{N} (a_j - b_j)^2,}$$

where $SE(p_1 - p_0)$ is the standard error of the difference. Miettinen (1969) has shown for large N that T has a standard normal distribution under the null hypothesis. He has also studied the power of the test and has given criteria, in terms of reducing cost, for deciding on an appropriate value of q, the number of comparison subjects per treatment subject.

When the outcome variable is continuous, one could compare the value for each treatment subject with the mean value of the corresponding controls, resulting in N differences. For a discussion of this approach, see Ury (1975).

Ury (1975) also presents an analysis of the statistical efficiency that can be gained by matching each case with several independent controls. For the dichotomous as well as the continuous outcome variables, the efficiency of using q controls versus a single control is approximately equal to $2q/(q + 1)$. For example, using 2 controls would increase the efficiency by about 33%; using 3 controls, by about 50%.

6.12 OTHER CONSIDERATIONS

This section includes three miscellaneous topics regarding matching. Sections 6.12.1 and 6.12.2 present results for matching that relate to general problems discussed in Sections 5.1 and 5.2, respectively: omitted confounding variables and measurement error. Some ideas regarding judging the quality of matches when exact matching is not possible are given in Section 6.12.3.

6.12.1 Omitted Confounding Variables

A common criticism investigators must face is that all the important confounding variables have not been taken into account. Unfortunately, with respect to matching, there are only very general indications of the effect of an omitted confounding variable.

Should the omitted confounding variable Z have a linear, parallel relationship with the included confounding variable X in the two populations, then matching solely on X removes only that part of the bias which can be attributed to the linear regression of Z on X. The amount of bias removed depends on the value of the regression coefficient of Z on X.

According to Cochran and Rubin (1973), if the regression of Z on X are nonlinear but parallel, then in large samples, matching solely on X will remove only that part of the bias due to Z that corresponds to the linear component of the regression of Z on X. These results generalize to the case of multiple confounding variables.

6.12.2   Errors of Measurement in Confounding Variables

If we assume that the response is linearly related to the correctly measured, or true, confounding variable in both populations, but that we can only match on values which are measured with error, then except under certain special conditions, the relationship between the response and fallible confounding variable will not be linear.

As an indication of the effect of measurement error on matching, consider the case where the response and the fallible confounding variable are linearly related. Then matching on the fallible variable has the effect of multiplying the expected percent reduction in bias by the ratio of $\beta*/\beta$ (Cochran and Rubin, 1973). In this ratio, $\beta*$ is the regression coefficient of the response on the fallible confounding variable and β is the regression coefficient of the response on the true confounding variable. Since this ratio is usually less than 1, matching on a confounding variable measured with error results in less bias reduction than does matching on the corresponding accurately measured confounding variable.

6.12.3   Quality of Pair Matches

In the case of pair matching the investigator can be lead to significant errors of interpretation if the quality of the matches is poor. Quality is judged by the magnitude of the differences between the values of the confounding variables for the comparison and treatment subjects. In this section we discuss a general approach that uses stratification to investigate any imperfect matching and its effect. We also discuss two approaches suggested by Yinger et al. (1967) for the case of a numerical outcome variable.

Perhaps the obvious first step in determining the overall quality of the pair matches obtained is to employ simple summary statistics such as the mean or median of the absolute differences between pairs for each particular confounding variable. Such statistics, however, do not give the investigator any indication of a relationship between the response and the closeness of matches. It is the existence of such a relationship which should be taken into account when interpreting the findings of a study. For example, if in a study on weight loss (numerical response) the pairs which show the greatest difference in weight loss were the pairs who were most imperfectly matched, the investigator should be suspicious of the apparent effect of the treatment.

How can investigators determine if there is any relationship between response and the quality of the matches? In the case of a categorical response, the investigators can take one of two approaches, depending on the number of confounding variables. If there are only a few variables, they can determine summary statistics for each response category. This may be viewed as analyzing the effect of possible imperfect matching by stratifying on the response. The summary statistics should be nearly equivalent for all response categories.

If there are several confounding variables, the investigators may instead wish to determine a single summary statistic of the quality of the matches for each response category. This can be done in a two-step procedure. First, for each confounding variable, the differences between matched pairs are categorized and weights are assigned to each category. For example, in the weight-loss study, if age is one of the confounding variables, a difference of 0 to 6 months may receive a weight of 0, and a difference of 7 to 11 months a weight of 1, while a difference of 12 months or more may receive a weight of 3. For the second step, the weights are summed across all confounding variables for each matched pair in a response strata and again we could either take the mean or the median as a summary statistic of the closeness of the matches. These numbers should agree across response strata. We wish to point out that these weights are arbitrary and are only meant to be used for within-study comparisons.

Yinger et al. (1967) have two methods for studying the effects of imperfect matching in the case of a numerical outcome. The first of their methods consists of forming a rough measure of the equivalence of the treatment and comparison groups by the weighting method discussed above for a categorical outcome. They call this measure the index of congruence.

Consider a study of reading ability, where age, sex, and birth order are confounding variables. Table 6.10 illustrates the calculation of the index of congruence for such a study. Here the index of congruence can range from 0 to 8 points, where a score of 0 indicates close matching and a score of 8 indicates the maximum possible difference between a treatment and control subject. Again, these scores are arbitrary and only meant as descriptive measures for within-study comparisons.

To determine if there is any relationship between the response and the quality of the matches, we can either calculate the correlation coefficient between the estimated treatment effect and the index of congruence, or plot the relationship. Ideally, both the correlation coefficient and the slope of the plotted curve should be close to zero, indicating no relationship.

The index of congruence gives only a rough measure of the group equivalence, in part because it does not take into account any directional influences of the confounding variables. The second of the Yinger et al. methods forms a directional measure of congruence which takes this factor into account.

The investigator may have prior knowledge (e.g., from previous research) of the directional influence of the confounding variables on the outcome. Con-

Table 6.10   Index of Congruence Calculation

Confounding Variable	Score	Range of Possible Point Differences between Matched Pairs
Age difference		
0–6 months	= 0	
7–11 months	= 1	0–3
12+ months	= 3	
Sex		
Same	= 0	0–3
Different	= 3	
Birth order		
Both either firstborn		
or not firstborn	= 0	0–2
Otherwise	= 2	
Total		0–8

sider again the study of reading ability and suppose that increasing age had a positive influence while increasing rank of birth had a negative influence. In addition, suppose the matching on sex was exact, so that we need not consider the directional influence. For the directional measure of congruence we will use only scores -1, 0, and 1, where -1 indicates that the treatment subject has a value of the confounding variable implying that the response for the treatment subject is expected to be inferior to that of the comparison subject; 0 indicates that they are expected to be the same, and 1 indicates that the treatment subject response is expected to be superior. Then, the directional index of congruence for matching a 12-year-old firstborn treatment subject with a 14-year-old second-born comparison subject of the same sex would be zero, since the treatment subjects superiority due to a lower birth order would be offset by his or her inferiority due to a lower age. In contrast, the index of congruence based on Table 6.10 for such a match would be 5.

The investigator may also plot the estimated treatment effect versus the value of the directional measure of congruence for each pair. By considering the scatter diagram and regression curve, the investigator can judge to what degree the treatment effect is related to differences between the matched pairs. Ideally, the plot should again show no relationship.

6.13   Conclusions

The most practical of the pair matching methods is nearest available matching. It has the advantage that matches can always be found. However, because of the varying tolerance, it will not be as effective as caliper matching in reducing the bias in the estimation of the treatment effect.

The pair matching methods are the best methods to use when the relationship is nonlinear. Rubin (1973a) found that the percentage reduction in bias for random-order nearest available matching in the linear case (Table 6.2) was overestimated by less than 10 percent in most nonlinear cases. Pair matching methods do require a large control reservoir, however, and are therefore difficult to use in studies with a large treatment group or where it takes a long time to find comparison subjects. They seem to be the most effective when σ_1^2/σ_0^2 is approximately 1 and to be least effective when σ_1^2/σ_0^2 is approximately 2 or more, with $\eta_1 > \eta_0$. Rubin also concludes that matching with $r \geq 2$ generally improves the estimate of the treatment effect, especially if the variance of the confounding variable is greater in the comparison population than in the treatment population.

Nonpair matching methods—mean and frequency—are quicker than are pair matching methods. However, mean matching is not used often because of its strong dependence on the assumption of linearity. If the investigator feels very confident that the relationship is a linear parallel one, and if the treatment and comparison groups are about the same size, mean matching may be considered as a fast matching procedure which has the same precision as pair matching in such a situation.

APPENDIX 6A: SOME MATHEMATICAL DETAILS

6A.1   Matching Model

The general mathematical model used to analyze the effect of matching with a numerical outcome variable can be represented as

$$Y_{ij} = R_i(X_{ij}) + e_{ij} \qquad i = 1, 0; j = 1, 2, \ldots, n_i, \qquad (6.13)$$

where $i = 1$ represents the treatment group, $i = 0$ the comparison group, and j is the jth observation in each group. Furthermore, Y_{ij} is the response variable and is a function of the confounding variable X_{ij}. The residual e_{ij} has mean zero and variance σ_i^2, and X_{ij} has mean η_i. We assume that the Y and X are numerical variables.

We now consider the specific forms of the response function $R_i(\cdot)$ that correspond to the linear parallel and quadratic parallel relationships.

6A.2   Parallel Linear Regression

For the parallel linear regression, the model (6.13) becomes

$$Y_{ij} = \mu_i + \beta(X_{ij} - \eta_i) + e_{ij} \qquad i = 1, 0; j = 1, 2, \ldots, n_i \qquad (6.14)$$

or

$$Y_{ij} = \alpha_i + \beta X_{ij} + e_{ij},$$

where

$$\alpha_i = \mu_i - \beta \eta_i.$$

Notice that the slope is the same for both the treatment and the comparison groups. In this case the *treatment effect* $\alpha_1 - \alpha_0$ is defined as the difference in the intercept terms:

$$\alpha_1 - \alpha_0 = (\mu_1 - \mu_0) - \beta(\eta_1 - \eta_0). \qquad (6.15)$$

Since the estimator of the treatment effect (6.15) is the difference in the mean responses between the two groups, $\overline{Y}_1 - \overline{Y}_0$, with mean response defined as $E(\overline{Y}_i | \overline{X}_i) = \alpha_i + \beta \overline{X}_i$, the expected value of the estimator is

$$E(\overline{Y}_1 - \overline{Y}_0 | \overline{X}_1, \overline{X}_0) = \alpha_1 - \alpha_0 + \beta(\overline{X}_1 - \overline{X}_0). \qquad (6.16)$$

From (6.16) it follows that the estimator $(\overline{Y}_1 - \overline{Y}_0)$ is biased by an amount $\beta(\overline{X}_1 - \overline{X}_0)$. Thus a matching procedure that makes $|\overline{X}_1 - \overline{X}_0|$ as small as possible will be preferred.

6A.3 Parallel Nonlinear Regression

Consider a parallel quadratic relationship. Then the matching model can be written as

$$Y_{ij} = \mu_i + \beta(X_{ij} - \eta_i) + \delta X_{ij}^2 + e_{ij}.$$

It follows that

$$E(\overline{Y}_1 - \overline{Y}_0 | \overline{X}_1, \overline{X}_0) = \mu_1 - \mu_0 - \beta(\eta_1 - \eta_0)$$
$$+ \beta(\overline{X}_1 - \overline{X}_0) + \delta(\overline{X}_1^2 - \overline{X}_0^2) + \delta(s_1^2 - s_0^2), \qquad (6.17)$$

where $s_i^2 = \sum_{j=1}^{N} (X_{ij} - \overline{X}_i)^2 / N$ for $i = 1, 0$. Comparing (6.17) with the treatment difference (6.15), we see that the bias equals

$$\beta(\overline{X}_1 - \overline{X}_0) + \delta(\overline{X}_1^2 - \overline{X}_0^2) + \delta(s_1^2 - s_0^2). \qquad (6.18)$$

Clearly, in this nonlinear case, equality of the confounding variable means is not sufficient. (In particular, mean matching is not appropriate.) This emphasizes the motivation of tight pair matching. By choosing pairs so that each treatment subject is very closely matched to a comparison subject, any difference in the sample confounding variable distributions that may be important [such as means and variances in (6.18)] is made small.

REFERENCES

Althauser, R. P., and Rubin, D. B. (1970), The Computerized Construction of a Matched Sample, *American Journal of Sociology*, **76**, 325-346.

Carpenter, R. G. (1977), Matching When Covariables Are Normally Distributed, *Biometrika*, **64**(2), 299—307.

Cochran, W. G. (1968), The Effectiveness of Adjustment by Subclassification in Removing Bias in Observational Studies, *Biometrics*, **24**(2), 295-313.

Cochran, W. G., and Rubin, D. B. (1973), Controlling Bias in Observational Studies: A Review, *Sankhyā, Series A*, **35**(4), 417-446.

Cox, D. R. (1957), Note on Grouping, *Journal of the American Statistical Association*, **52**(280), 543-547.

Fleiss, J. L. (1973), *Statistical Methods for Rates and Proportions*, New York: Wiley.

Greenberg, B. G. (1953), The Use of Analysis of Covariance and Balancing in Analytical Surveys, *American Journal of Public Health, Part I*, **43**(6) 692-699.

Haddon, W., Jr., Valien, P., McCarroll, J. R., and Umberger, C. J. (1961), A Controlled Investigation of the Characteristics of Adult Pedestrians Fatally Injured by Motor Vehicles in Manhattan, *Journal of Chronic Disease*, **14**, 655-678. Reprinted in E. R. Tufte, Ed., *The Quantitative Analysis of Social Problems*, Reading, MA: Addison-Wesley, 1970, pp. 126-152.

Kalton, G. (1968), Standardization: A Technique to Control for Extraneous Variables, *Journal of the Royal Statistical Society, Series C*, **17**, 118-136.

Lachenbruch, P. A. (1979), Discriminant Analysis, *Biometrics*, **35**(1), 69-85.

McKinlay, S. M. (1974), The Expected Number of Matches and Its Variance for Matched-Pair Designs, *Journal of the Royal Statistical Society, Series C*, **23**(3), 372-383.

McKinlay, S. M. (1975), The Effect of Bias on Estimators of Relative Risk for Pair-Matched and Stratified Samples, *Journal of the American Statistical Association*, **70**(352), 859-864.

McKinlay, S. M. (1978), The Effect of Nonzero Second-Order Interaction on Combined Estimators of the Odds Ratio, *Biometrika*, **65**, 191-202.

Mantel, N., and Haenszel, W. (1959), Statistical Aspects of the the Analysis of Data from Retrospective Studies of Disease, *Journal of the National Cancer Institute*, **22**, 719-748.

Miettinen, O. S. (1969), Individual Matching with Multiple Controls in the Case of All-or-None Responses, *Biometrics*, **25**(2), 339-355.

Miettinen, O. S. (1970), Estimation of Relative Risk from Individually Matched Series, *Biometrics*, **26**(1), 75-86.

Rubin, D. B. (1973a), Matching to Remove Bias in Observational Studies, *Biometrics*, **29**(1), 159-183.

Rubin, D. B. (1973b), The Use of Matched Sampling and Regression Adjustment to Remove Bias in Observational Studies, *Biometrics*, **29**(1), 185-203.

Rubin, D. B. (1976a), Multivariate Matching Methods That Are Equal Percent Bias Reducing, I: Some Examples, *Biometrics*, **32**(1), 109-120; 955.

Rubin, D. B. (1976b), Multivariate Matching Methods That Are Equal Percent Bias Reducing, II: Maximums on Bias Reduction for Fixed Samples Sizes, *Biometrics,* **32**(1), 121–132, 955.

Rubin, D. B. (1979), Using Multivariate Matched Sampling and Regression Adjustment to Control Bias in Observational Studies, *Journal of the American Statistical Association,* **74,** 318–328.

Smith, A. H., Kark, J. D., Cassel, J. C., and Spears, G. F. S. (1977), Analysis of Prospective Epidemiologic Studies by Minimum Distance Case-Control Matching, *American Journal of Epidemiology,* **105**(6), 567–574.

Snedecor, G. W., and Cochran, W. G. (1967), *Statistical Methods,* 6th ed., Ames, IA: Iowa State University Press.

Ury, H. K. (1975), Efficiency of Case-Control Studies with Multiple Controls per Case: Continuous or Dichotomous Data, *Biometrics,* **31**, 643–649.

Yinger, J. M., Ikeda, K., and Laycock, F. (1967), Treating Matching as a Variable in a Sociological Experiment, *American Sociological Review,* **23**, 801–812.

CHAPTER 7

Standardization and
Stratification

Standardization and stratification are related adjustment procedures that are applicable when the confounding factor is categorical. The outcome and risk factors may be either categorical or numerical, except that for application to cohort studies, the risk factor must be categorical, and for case-control studies, the outcome must be categorical.

The goal of any adjustment procedure is to correct for differences in the confounding factor distributions between the treatment groups. Standardization does this by estimating what would have been observed had the confounding factor distributions been the same in the two groups being compared. In the data to be presented in Section 7.1, for example, the rates of death due to breast cancer are compared for two groups of women. The two groups of women differ with respect to age, an important confounding factor when comparing death rates. Standardization in this case estimates death rates based on a common age distribution. This common age distribution, or more generally, the common confounding factor distribution is taken from some other group, known as the standard population; hence the term *"standard*ization."

The term "stratification," when applied to adjustment procedures, can be used in two ways. The first and more general use is to describe any adjustment procedure that divides the study population into groups (strata) based on the values of the confounding factors and then combines information across groups to provide an estimate of the treatment effect. In this general sense, standardization is a stratified procedure where the standard population provides the basis for combining information across strata.

In this book, however, the terms "stratification" or "stratified analysis" will only be used in a second, more restrictive, way. This second usage is consistent with the first in that the study population is divided into strata and information is combined across groups. The restriction is that the basis for combining across groups be some statistical criterion, such as maximum likelihood or minimum variance, without reference to any standard population.

Standardization and stratification are employed for two purposes: (*a*) to provide summary statistics for comparing different populations with respect to such items as mortality, price levels, or accident rates; and (*b*) to yield estimators of the difference in rates or means between two populations or of the relative risk (θ) or odds ratio (ψ) that are unbiased, or at least approximately so.

In Section 7.1 we present the principles of standardization for the simple case of a cohort study with dichotomous risk and outcome factors. Some considerations in the choice of a standard population and standardization procedure are given in Sections 7.2 and 7.3, respectively; the bias and precision of standardized estimators are discussed in Section 7.4; and the extension to case-control studies is considered in Section 7.5. General formulas for direct and indirect standardization and more detailed bias considerations are given in Appendix 7A.

Stratification is introduced in Section 7.6 with emphasis on estimators of the odds ratio. The odds ratio estimators are considered in greater mathematical detail in Appendix 7B. The extension of standardization and stratification to numerical outcomes and multiple confounding factors is presented in Sections 7.7 and 7.8.

If the confounding variable is numerical, standardization and stratification can be applied by first categorizing the confounding variable (as in Table 7.1 for the numerical confounding variable, age). The effects of this categorization on the bias of the stratified estimators are discussed in Chapter 13. For now, it is sufficient to note that the estimators will always be biased, even in large samples. Comments in this chapter on bias are for a categorical confounding variable (except when considering McKinlay's work in Section 7.6.2). For the case of a numerical confounding variable, Cochran's work (1968) for frequency matching (Section 6.7) gives some guidance for choosing the number and sizes of the strata. Logit analysis (Chapter 9) and analysis of covariance (Chapter 8) are alternative procedures that do not require stratifying a numerical confounding variable.

7.1 STANDARDIZATION—EXAMPLE AND BASIC INFORMATION

The data in Table 7.1 on breast cancer death rates among females aged 25 or older is based on work by Herring (1936, Tables I and II). In this case, the risk factor is marital status with two categories, single (never married) and married (including widows and divorcees), and the outcome is death due to breast cancer. The breast cancer death rates, per 100,000 population, of 15.2 for single women and 32.3 for women who were ever married are called *crude* (or unadjusted) *death rates* because they have not been adjusted for any possible confounding. Crude rates are calculated by simply ignoring any possible confounding factors. For example, the crude rate for single women is found by dividing the average annual number of deaths due to breast cancer (1438) by the total population of single women (94.73). The crude relative risk of death due to breast cancer is 2.1 = 32.3 ÷ 15.2, purporting to indicate that women who marry are more than twice as likely to die of breast cancer as are women who never marry.

An examination of the death rates by age—the *age-specific rates*—indicates that age is an important factor. Age is a confounding factor in this circumstance because death rates increase with age and the ages of married and single women differ; the married women tend to be older (80% of the single women are younger than 35 as compared to only 35% of the women who had married). (The definition of a confounding factor is given in Section 2.1.) A fair comparison of the cancer death rates requires that age, at least, be adjusted for.

One approach to adjustment is *direct standardization*. This approach asks

Table 7.1   Breast Cancer Mortality in Females in the United States (1929–1931)

Age (yr)	Single Women		Ever-Married Women		All Women	
	1930 Population (100,000's)	Average Annual Breast Cancer death rate (per 100,000 population)	1930 Population (100,000's)	Average Annual Breast Cancer death rate (per 100,000 population)	1930 Population (100,000's)	Average Annual Breast Cancer death rate (per 100,000 population)
15–34	76.15	0.6	89.57	2.5	165.72	1.6
35–44	7.59	24.9	61.65	17.9	69.24	18.7
45–54	5.22	74.7	46.67	43.1	51.89	46.2
55–64	3.43	119.7	31.11	70.7	34.54	75.5
65–74	1.88	139.4	18.14	89.4	20.02	94.1
≥75	0.45	303.8	7.80	137.7	8.25	146.8
Total	94.73	15.2	254.94	32.3	349.67	27.6
Average annual Number of deaths	1438		8228		9666	

how many cancer deaths would have occurred if the age distribution for both the married and single women had been the same as in some standard population, but the age-specific rates were the same as observed? In this example, a natural standard is the 1930 age distribution of all women in the United States. The directly standardized cancer mortality rates are (per 100,000 population):

Single women:

$$\frac{(165.72)\,(0.6) + \cdots + (8.25)\,(303.8)}{349.67} = \frac{15,131.27}{349.67} = 43.3$$

Married women:

$$\frac{(165.72)\,(2.5) + \cdots + (8.25)\,(137.7)}{349.67} = \frac{9257.95}{349.67} = 26.5.$$

Single women have a higher age-adjusted mortality rate, and the directly standardized relative risk is $\hat{\theta}^D = 26.5/43.3 = 0.6$, indicating that, after correcting for age differences, women who marry actually have a lower risk of dying of breast cancer than do women who remain single. Note that $\hat{\theta}^D$ is the ratio of the expected numbers of cancer deaths in the standard population based on the age-specific death rates for married women (9257.95) to the expected number of deaths in the standard population based on the rates for single women (15,131.27).

An alternative approach is *indirect standardization.* This approach asks how many cancer deaths would have occurred among single women if the age distributions for the single and married women were the same as observed, but the age-specific mortality rates had been the same as in some standard population? As we will see in Section 7.3, when applying indirect standardization, it is best to select the "standard population" to be one of the two groups being compared.

The results of indirect standardization are often quoted as *standard mortality* (or morbidity) *ratios,* which are ratios of the observed deaths for each category of the risk factor to the expected deaths given the standard age-specific rates. The indirectly standardized rate for each risk factor category is then found by multiplying the standard mortality ratio for that category by the crude rate for the standard population. (The rationale for this round about calculation of a standardized rate is given in Section 7A.2.) Indirectly standardized relative risks, $\hat{\theta}^I$, are found as ratios of the indirectly standardized rates. In the special case where the standard rates are taken to be those corresponding to one of the two groups being compared, the standard mortality ratio itself turns out to be a relative risk.

For our example, taking the standard rates to be the age-specific breast cancer death rates for married women, the expected deaths for single women are

$$(76.15)\,(2.5) + \cdots + (0.45)\,(137.7) = 1023.8.$$

The standard mortality ratio is then $1.40 = 1438 \div 1023.8$, indicating that there were 40% more deaths among single women than would have been expected if their age-specific rates had been the same as for married women. The indirectly standardized breast cancer death rate for single women is found by multiplying the standard mortality ratio by the crude mortality rate for the married women:

$$(1.40)(32.3) = 45.2.$$

The indirectly standardized relative risk comparing married to single women is then

$$\hat{\theta}^I = \frac{32.3}{45.2} = \frac{1}{1.40} = 0.71.$$

The standard mortality ratio of 1.40 is the inverse relative risk, that is, of remaining single compared to getting married.

It is common for different standardization procedures to yield different standardized rates and estimates of relative risk as occurred here. This emphasizes that the primary purpose of standardization is to provide a single summary statistic (mortality rates here) for each category of the risk factor so that the categories may be compared. The numbers that are most meaningful are the age-specific mortality rates, and standardization is not a substitute for reporting the specific rates.

To emphasize this point further, look again at the specific rates in Table 7.1. For women under the age of 35, married women have a slightly higher breast cancer death rate than do single women; for women age 35 or over, single women have the higher breast cancer mortality rate, and the difference and ratio of the rates varies with age. This is an example of interaction between the confounding factor age and the treatment effect (Section 3.3). This interaction is an important fact that is not conveyed by the reporting of a single relative risk, standardized or not. One consequence of this is that a different choice of a standard population could have resulted in standardized rates that, like the crude rates, were higher for married women than for single. The choice of standard thus becomes an important issue. Some guidelines for choosing the standard will be discussed in Section 7.2. Spiegelman and Marks (1966) and Keyfitz (1966) give examples of how different choices for the standard population can affect the standardized rates and relative risks. Keyfitz compares the 1963 female mortality rates in 11 countries using three different standard age distributions and finds that the ranking of the countries depends on the choice of standard.

The direct and indirect methods are the most important standardization procedures, but not the only ones. There are many standardized indices that have been developed for particular fields. Kitagawa (1964, 1966) discusses many of these alternatives, particularly with reference to demography.

7.2 CHOICE OF STANDARD POPULATION

The choice of standard population is, in general, a contextual decision. When standardization is being employed for comparison purposes, there are two commonsense guidelines for choosing the standard population. The first is to use the data for the entire population that the study subjects are chosen from. This was done for the breast cancer mortality example by using the data for all women in the United States in 1930 as the standard for direct standardization. If the population data are not available, an alternative is to combine all the risk factor groups (i.e., take the standard population to be the entire sample being studied). The rationale behind this alternative is that summing over the risk factor should yield a "population" that approximates the real population of interest. This approach will approximate the population well if the sampling fractions in the risk factor groups are equal, or nearly so; the approximation will be poor, if, for example, the subjects in one category of the risk factor are all persons known to have been exposed to the risk factor and the subjects in the second category are only a portion of those not exposed.

The second guideline is applicable in cases where all but one category of the risk factor correspond to a treatment of some sort, and the remaining category corresponds to the absence of a treatment. Then, the nontreatment category is a reasonable choice of standard population. For example, when Cochran (1968) standardized lung cancer rates for age, he chose the nonsmokers to be the standard population. The cigarette smokers and cigar and/or pipe smokers were two "treatment" groups.

As discussed in Section 7.1, the choice of standard can make a difference in the comparison of risk factor groups. Therefore, it is important to report the specific rates. If a single summary statistic is still required, the standard should be picked to resemble the risk factor groups as much as possible, so as to preserve, to the extent possible, the meaningfulness of the standardized comparison. For the data of Table 7.1, for example, the 1960 age distribution of males in Mexico would be an inappropriate choice of standard.

7.3 CHOICE OF STANDARDIZATION PROCEDURE

The choice of standardization procedure can depend on many considerations. If the total sample size and specific rates for the categories of the risk factor are known but the numbers of individuals at each level of the confounding factor are not known, then directly standardized rates can be calculated but indirectly standardized rates cannot. Conversely, indirectly standardized rates can be calculated if the specific rates are not known but the total number of deaths (outcomes) and the specific rates for the standard are known.

There is one important caution regarding indirect standardization. It is possible to have two categories of the risk factor with identical specific rates but different indirectly standardized rates. To see this, consider the artificial data in Table 7.2, with three risk factor groups and two categories in the confounding factor. The specific rates for the first two risk factor categories are identical so a proper adjustment procedure should yield a relative risk of 1.0. The crude rates are 0.82 for the first category and 0.18 for the second, so the crude relative risk is 4.56, reflecting the very different confounding factor distributions in the two groups being compared.

Now consider the direct and indirect standardized relative risks with the total of the three risk factor groups as the standard population. Following the procedures of Section 7.1, the directly standardized rates for the first two risk factor categories are both 0.50, so the directly standardized relative risk is 1.0. The indirectly standardized rates are 0.62 for the first risk factor category and 0.26

Table 7.2   Artificial Data to Demonstrate Comparability Problem of
Indirect Standardization

Confounding Factor Category	Risk Factor Category							
	1		2		3		Total	
	Sample Size	Rate	Sample Size	Rate	Sample Size	Rate	Sample Size	Rate
1	900	0.9	100	0.9	1000	0.5	2000	0.7
2	100	0.1	900	0.1	1000	0.5	2000	0.3

for the second, yielding an indirectly standardized relative risk of 2.38. This result is more reasonable than the crude relative risk of 4.56, but still not good. The indirect method would only have worked in this example if the specific rates in the standard population happened to be the same as for categories 1 and 2 of the risk factor.

Indirect standardization is best used only for comparing two groups when one of those groups is the standard. In that case, the two methods of standardization are equivalent, in the sense that equal estimates of θ can be obtained by particular choices of the standard for each method. In addition, the indirectly standardized rates will be equal if all the specific rates are equal. Mathematical details are given in Sections 7A.3 and 7A.4.

7.4   STATISTICAL CONSIDERATIONS FOR STANDARDIZATION

As with other adjustment techniques, standardized estimation of treatment effects is most meaningful when the treatment effect is constant over the confounding factor strata [i.e., when there is no interaction (Section 3.3)]. In this section we will consider the bias and precision (variance) of the standardized estimators of the constant treatment effect, whether relative risk or difference of rates.

7.4.1   Bias

If the difference in risk factor rates is the same for each category of the confounding factor, direct standardization yields unbiased estimates of the difference between the risk factor rates. For estimating the relative risk, if the sample relative risks are the same, say $\hat{\theta}$, within each category of the confounding factor, then the directly standardized estimate of relative risk is also equal to θ (as demonstrated in Section 7A.3.1). This implies that the directly standardized relative risk will be approximately unbiased in large samples.

In general, the indirectly standardized estimators of both parameters are biased. An exception occurs when comparing two groups by means of relative risk and one of these groups is the standard (see Section 7A.4). Indirect standardization does not yield unbiased estimates of the difference between the rates because the standard mortality ratio is a ratio.

7.4.2 Precision

Indirect standardization has been found to be more precise than direct standardization for estimating rates (Bishop, 1967) and relative risks (Goldman, 1971). Goldman further showed that the precision of directly standardized relative risks could be improved by first applying the log-linear model technique (Chapter 10) and then directly standardizing using the fitted rates. Details are presented in Bishop (1967), Goldman (1971), and Bishop et al. (1974, Sec. 4.3).

7.5 EXTENSION OF STANDARDIZATION TO CASE-CONTROL STUDIES

Since we cannot estimate rates directly (see Section 3.1), much of the previous material is not applicable to case-control studies. We must instead ask how to obtain a standardized estimate of the odds ratio. Miettinen (1972) developed a procedure motivated by the idea, presented in Section 7.1, that the standardized relative risks are ratios of expected numbers of deaths (or other dichotomous outcomes). For case-control studies, Miettinen proposed using the ratio of the expected numbers of cases in the two risk factor groups, where the expectation is based on a standard distribution of numbers of controls. Letting C_k, a_{rk}, and c_{rk} denote standard numbers of controls, observed numbers of cases, and observed numbers of controls, respectively, where k denotes the categories of the confounding factor and r the categories of the risk factor ($r = 1$ if the risk factor is present; $r = 0$ if not), the standardized estimator of the odds ratio is

$$\hat{\psi}^M = \frac{\sum\limits_{k=1}^{K} C_k(a_{1k}/c_{1k})}{\sum\limits_{k=1}^{K} C_k(a_{0k}/c_{0k})}.$$

To see that $\hat{\psi}^M$ is the ratio of the "expected" number of cases in the two risk factor groups, let A_{1k} be the number of cases corresponding to C_k controls in the risk factor present group. To find A_{1k}, set the ratio of expected numbers of cases to controls equal to the observed ratio

$$\frac{A_{1k}}{C_k} = \frac{a_{1k}}{c_{1k}}$$

and solve for A_{1k}:

$$A_{1k} = \frac{C_k a_{1k}}{c_{1k}}.$$

Similarly,

$$A_{0k} = \frac{C_k a_{0k}}{c_{0k}}.$$

Summing A_{1k} over the K confounding factor strata yields the total number of expected cases, the numerator of $\hat{\psi}^M$. The ratio $\hat{\psi}^M$ then compares the number of cases expected in the risk-factor-present group to the number expected in the risk-factor-absent group, based on the same standard distribution of controls (the C_k). Considerations in the choice of standard, discussed in Section 7.2, apply here as well.

Miettinen shows that $\hat{\psi}^M$ can be written as a weighted average of the odds ratios from each of the K confounding factor strata (the specific odds ratios). Therefore, if the odds ratios, ψ_k, are constant over all categories of the confounding factor, $\hat{\psi}^M$ will be an approximately unbiased estimator of ψ in large samples (within each stratum).

7.6 STRATIFICATION

The method of stratification differs from standardization in that a statistical criterion, such as minimum variance, rather than a standard population, is the basis for combining across confounding factor strata. In this section we will cover the best-studied case of stratification, that of estimating the odds ratio for dichotomous risk and outcome factors in either cohort or case-control studies. Throughout Section 7.6 it will be assumed that the odds ratio is the same in all the confounding factor strata. The formulas for the various estimators are presented in Appendix 7B. Stratified estimation of the difference of means is covered in Section 7.7

7.6.1 Estimators of the Odds Ratio

A large number of estimators of the odds ratio have been proposed. Gart (1962) presented the maximum likelihood estimator and (1966) a modification to Woolf's (1955) estimator (the "modified Woolf" estimator). Birch's (1964) and Gart's (1970) estimators are approximations to yet another estimator, the conditional maximum likelihood estimator (Gart, 1970). Goodman (1969) proposed approximations to the maximum likelihood and conditional maximum likelihood estimators. A well-known estimator is that of Mantel and Haenszel (1959).

There are two different maximum likelihood estimators, usual and conditional, because there are two different sampling situations to be considered. As the theoretical properties of the various estimators depend on which sampling situation is appropriate, we must begin with an explanation of the two cases, specifically emphasizing what is meant by a large sample in each of the two cases.

Consider the situation where, within each confounding factor stratum there is some number of subjects in each of the two study groups, and suppose that we wish to add more subjects so as to increase the total sample size. Then, there are two choices: more subjects can be added to the existing strata; or new strata can be added with corresponding, additional subjects.

The first case is the most commonly considered. Often the number of strata is fixed by the nature of the situation. For example, if the confounding factor was sex, the number of strata are fixed at two and the sample size can be increased only by adding more males and females. In such a situation, "large sample" means that the sample sizes in each study group within each strata are large, regardless of the number of strata.

Consider now a study that is conducted cooperatively in many institutions, and suppose that institution is the confounding variable of interest. Each stratum will then consist of the subjects from a particular institution. In this study a larger sample could be obtained in two ways. The first would be as above, namely adding subjects from each currently participating institution. The second is to add more institutions and select subjects from the new institutions. For every additional institution there will be an additional stratum for the confounding factor. "Large sample" in this second case means that the number of confounding factor strata is large, regardless of the sample sizes within each stratum.

To summarize, there are two definitions of large sample. In the first, the sample sizes within each stratum are large; in the second, the number of strata are large. This distinction is important because estimators can behave differently in the two cases. In particular, the properties of the (usual) maximum likelihood estimator apply only in the first case. The second case requires a different estimator, the conditional maximum likelihood estimator. Each maximum likelihood estimator will be approximately unbiased and normally distributed in large samples as defined for the appropriate sampling scheme (Gart, 1962; Andersen, 1970).

The numerical difficulties in solving for the conditional maximum likelihood estimator led Birch (1964) and Goodman (1969) to propose approximations that are easier to calculate. The Birch and Goodman estimators are unbiased in large samples only if the odds ratio is 1. In terms of bias considerations, the conditional maximum likelihood estimator is therefore preferable. Gart's (1970) approximation to the conditional maximum likelihood estimator is applicable if the sample sizes within each stratum are large. This approximation will be approximately unbiased when both the number of strata and sample sizes within each stratum are large. Although the unbiasedness holds for any value of the odds ratio, requiring both the number of strata and the sample sizes to be large is very restrictive.

The (usual) maximum likelihood estimator that is appropriate when the sample sizes are large within each stratum is the basis of comparison for the remaining estimators. In the appropriate large samples, this estimator is approximately unbiased and no other unbiased estimator has a lower variance. In large samples, then, this is a good choice of estimator.

Woolf's (1955) estimator is equivalent, in the sense of having the same large-sample distribution, to the maximum likelihood estimator. However, as will be shown in detail in Section 7B.2, this estimator cannot be calculated if either of the observed proportions in *any* stratum is 0 or 1. In practice this means the Woolf estimator will be virtually useless when dealing with rare outcomes in cohort studies or rare risk factors in case-control studies.

Gart (1966) suggested a modification to the Woolf estimator that does not suffer from this problem while retaining the large-sample equivalence to the maximum likelihood estimator. There are thus three estimators that are equivalent in large samples: the maximum likelihood, Woolf, and modified Woolf Estimators. What is known about their small-sample properties is considered in Section 7.6.2.

The Mantel–Haenszel (1959) estimator is also approximately unbiased and normally distributed in large samples (large within each stratum), but its variance is larger than that of the maximum likelihood estimator unless the odds ratio is 1 (Hauck, 1979). In large samples, then, one of the three equivalent estimators noted above would be preferable to the Mantel–Haenszel estimator.

7.6.2 Comparisons of Odds Ratio Estimators

Using simulation, McKinlay (1975) compared the bias, precision, and mean squared error of the modified Woolf, Mantel–Haenszel, and Birch estimators for the case of a numerical confounding variable that is stratified into various numbers of strata. As mentioned earlier, all the standardized and stratified estimators will be biased in such a case, even in large samples. McKinlay's work is discussed in greater detail in Section 12.2.2.

Based on the mean squared errors of the estimators, McKinlay recommended the modified Woolf estimator, but with some reservations. The modified Woolf estimator has a smaller variance than the Mantel–Haenszel estimator, and this is reflected in the mean squared errors. However, the bias of the modified Woolf estimator increases with increasing number of strata. McKinlay noted that "only Mantel and Haenszel's estimator consistently removed bias in all the simulated situations considered—a property which is masked in this investigation by the relatively large variance" (p. 863). In terms of bias removal, then, the Mantel–Haenszel estimator is to be preferred. In addition, the difference in mean squared errors between the Mantel–Haenszel and modified Woolf estimators became negligible for the large samples (total sample size of 600) in McKinlay's study.

In an unpublished study, W. Hauck, F. Leahy, and S. Anderson addressed the question of whether McKinlay's 1975 results are applicable to the case of a categorical confounding factor where the estimators would be approximately unbiased in large samples. This study was patterned after McKinlay's 1975 study and compared the modified Woolf, Mantel–Haenszel, and (usual) maximum likelihood estimators. In terms of bias, the modified Woolf was least and maximum likelihood most biased, except for increasing values of the odds ratio and large sample sizes where the Mantel–Haenszel estimator was the least biased and the modified Woolf the most. However, the bias was small for all three estimators for the sample sizes and number of strata considered by McKinlay. In terms of variance and mean squared error, the modified Woolf was more precise, sometimes considerably so, than the other two and the maximum likelihood estimator least precise.

In another simulation study with a numerical confounding factor and large samples (total sample sizes of 200 to 1000), McKinlay (1978) compared the Mantel–Haenszel estimator, Gart's (1970) asymptotic approximation to the conditional maximum likelihood estimator, and the modified Woolf estimator. The results for the modified Woolf and Mantel–Haenszel estimators were similar to her 1975 results, namely that the Woolf estimator was usually most precise but that its bias tended to increase with increasing numbers of strata from 2 to 10, while the Mantel–Haenszel estimator was preferable in terms of bias, particularly for the larger number of strata. The Gart estimator, a close approximation to the conditional maximum likelihood estimator in the cases considered by McKinlay, was never better than the Mantel–Haenszel estimator in terms of either bias or precision.

The three studies agree that on purely bias considerations, the Mantel–Haenszel estimator is best, selected over the modified Woolf estimator on the grounds of consistency. If precision is taken into account by considering mean squared error, then, for the cases considered, the modified Woolf estimator is best.

This is an example of a common statistical problem of making a trade-off between bias and precision. Since the modified Woolf estimator is sometimes considerably more precise, and since the biases of all the estimators considered are not large, this would seem to be the estimator of choice. What is of concern, however, is the tendency for the bias of the modified Woolf estimator to increase with increasing number of strata for a given total sample size and with increasing distance of the odds ratio from the null value of 1. This implies that the modified Woolf estimator is more sensitive to the sample sizes within each stratum than is the Mantel–Haenszel estimator. Consequently, the modified Woolf estimator can be clearly preferred only for a small number of strata with large sample sizes within each stratum; otherwise, the Mantel–Haenszel estimator is a good choice.

7.7 STANDARDIZATION AND STRATIFICATION FOR NUMERICAL OUTCOME VARIABLES

The standardization results of Sections 7.1 to 7.3 and Appendix 7A apply to numerical outcome variables with mean responses replacing the rates. The principal difference in using a numerical outcome is that interest generally shifts to differences, such as the difference in means, instead of ratios, such as the relative risk. If the mean treatment difference $\tau = \alpha_1 - \alpha_0$ is constant over all levels of the confounding factor, direct standardization will yield unbiased estimates of the treatment effect. As with the difference of rates, indirect standardization yields biased estimates.

Stratified estimators of the mean treatment difference have the form of a weighted combination of the difference of means within each of the confounding factor strata:

$$\frac{\sum_{k=1}^{K} v_k (\overline{Y}_{1k} - \overline{Y}_{0k})}{\sum_{k=1}^{K} v_k}.$$

To minimize the variance of the stratified estimator, the weights, the v_k, are chosen, where possible, inversely proportional to the variance of $\overline{Y}_{1k} - \overline{Y}_{0k}$. In the simplest case, the variance of each observation is constant in both risk factor groups and in all confounding factor strata. Then

$$v_k = \left(\frac{1}{n_{0k}} + \frac{1}{n_{1k}} \right)^{-1}. \tag{7.1}$$

Kalton (1968) discusses the choice of the weights in detail, including the use of estimated variances. If the weights are constants, such as in (7.1), the stratified estimator will be unbiased. If the weights depend on the sample data, as by the use of estimated variances, the stratified estimators are biased, but the bias will become negligible in large samples.

7.8 EXTENSION TO MORE THAN ONE CONFOUNDING FACTOR

More than one categorical confounding factor can easily be handled by treating them together as one confounding factor. For example, two dichotomous confounding factors can be combined to form a single four-category confounding factor. However, as the number of confounding factors increases, the number of categories in the combined confounding factor can grow very quickly. This leads to the problem of small numbers in each category of the confounding factor and consequently, poorly determined specific rates or means.

There are really only two solutions to this problem. The first is to be selective in choosing confounding factors to adjust for. The second is to first apply the log-linear model technique (see Section 7.4.2). An extreme situation is that the number of categories in the combined confounding factor may be so large that some of the sample sizes on which the specific rates would be based are zero. Direct standardization cannot then be applied. Application of log-linear analysis eliminates the zeros.

Indirect standardization is frequently advocated because it is more precise than direct standardization, particularly in the presence of small numbers. This is true because indirect standardization does not use the specific rates, which will be poorly determined in small samples. As elaborated upon earlier, the general use of indirect standardization is not recommended. However, if there are too many confounding factor categories and many zeros, precluding the use of direct standardization, indirect standardization can still be applied and be better than the crude rates.

For purposes of stratification, Miettinen (1976) has proposed a method for reducing a set of confounding factors, whether numerical or discrete, to a single numerical confounding factor. The resulting confounding factor, "confounder score" in Miettinen's terminology, can then be categorized and the procedures of Section 7.1 or 7.6 applied. This method may be applied to either cohort or case-control studies, as long as both the outcome and risk factors are dichotomous.

The basis of Miettinen's proposal, as for discriminant matching (Section 6.10.4), is to use a discriminant function to distinguish (discriminate) between cases and noncases. This discriminant function depends on the value of the risk factor and the confounding factors. (This is stated for cohort studies; for case-control studies, distinguish instead between the risk-factor-present and risk-factor-absent groups. The discriminant function will then depend on the values of the outcome and the confounding factors.) The confounder score for each individual is obtained by evaluating the discriminant function for that person, assuming that the person is in the risk-factor-absent group, regardless of which risk factor group he or she is actually in. (For case-control studies, the function is evaluated assuming the person to be a control.) The motivation for this method is that the confounder score is a single variable that may be interpreted as a risk score that takes into account all variables except the risk factor.

7.9 HYPOTHESIS TESTING

In conjunction with the estimation problem, it is frequently desired to test the hypothesis that the risk factor has no effect on the outcome. Tests of $\theta = 1$ based on standardized relative risks can be done by using a standard normal distribution test. The necessary standard error formulas are given by Chiang (1961) and Keyfitz (1966).

Tests for the odds ratio related to stratified estimators are due to Mantel and Haenszel (1959) and Gart (1962), the latter being related to Woolf's estimator, and for the difference of rates due to Cochran (1954). The odds ratio procedures allow us to test whether the odds ratio is the same in all confounding factor strata (i.e., test whether the no interaction assumption is valid), and then whether the common value of the odds ratio differs from 1. Cochran's procedure does the same for the difference of rates, testing whether the common value of the difference differs from zero. Alternatives are likelihood ratio tests in log-linear analysis (Chapter 10). Much of this material is reviewed by Gart (1971), whose paper contains an extensive bibliography, and by Fleiss (1973, Chap. 10).

APPENDIX 7A MATHEMATICAL DETAILS OF STANDARDIZATION

At this point, the assumption of a dichotomous risk factor will be loosened to allow a general categorical risk factor. It will still be assumed that the response is dichotomous, so that the discussion will be in terms of rates, and that the data were obtained from a cohort study. No assumption is made regarding the choice of the standard population. It may, for example, correspond to one of the categories of the risk factor.

7A.1 Notation

Let R and K denote the number of categories in the risk factor and confounding factor, respectively. Lowercase letters, r and k, will be used as the corresponding indices. Let p_{rk} denote the observed rate based on n_{rk} individuals for the rth category of the risk factor and the kth category of the confounding factor. For the example in Table 7.1, there are $R = 2$ categories of the risk factor marital status, $K = 6$ categories of the confounding factor age, and, for example, $n_{11} = 76.15 \times 10^5$ women in the group corresponding to the first category of marital status and the first category of age, and $p_{26} = 137.7 \times 10^{-5}$ is the breast cancer death rate for women in the second marital status category and sixth age category. The standard population has N_k individuals in the kth category of the confounding factor, with a corresponding rate of P_k. If an index is replaced by a dot, it indicates summation over that index. For example,

$$n_{r.} = \sum_{k=1}^{K} n_{rk}.$$

The crude rate for the rth category of the risk factor is then found as

$$p_r^C = \frac{1}{n_{r.}} \sum_{k=1}^{K} n_{rk}p_{rk},$$

and for the standard population it is

$$P = \frac{1}{N.} \sum_{k=1}^{K} N_k P_k.$$

7A.2   Computation of Directly and Indirectly Standardized Rates

The directly standardized rate for the rth category of the risk factor is

$$p_r^D = \frac{1}{N.} \sum_{k=1}^{K} N_k p_{rk}.$$

The standard mortality ratio (SMR) is

$$\mathrm{SMR}_r = \frac{\sum\limits_{k=1}^{K} n_{rk}p_{rk}}{\sum\limits_{k=1}^{K} n_{rk}P_k} = \frac{p_r^C}{(1/n_{r.}) \sum\limits_{k=1}^{K} n_{rk}P_k},$$

and the indirectly standardized rate is

$$p_r^I = \mathrm{SMR}_r \times P.$$

We can see here why the roundabout calculation of indirectly standardized rates, beginning with the computation of the standard mortality ratio, is necessary. The straightforward analog of direct standardization would be to use

$$\frac{1}{n_{r.}} \sum_{k=1}^{K} n_{rk}P_k$$

as the indirectly standardized rate. This, however, is the rate for the standard population directly standardized to the confounding factor distribution in the rth category of the risk factor and so is not a rate that reflects the influence of the risk factor category. The standard mortality ratio acts as a correction factor to the standard population rate, P. The standard mortality ratio is the ratio of observed to expected deaths in the rth risk factor group, where the expectation is with respect to the standard population specific rates. The standard mortality ratio, then, indicates how much P should be changed to reflect the specific rates in the risk factor group.

7A.3   Bias of Indirect Standardization

The bias and consequent interpretability problems of indirect standardization are sufficiently important to be elaborated further. Estimation of the relative risk, θ, and difference of rates, Δ, are considered separately. In Section 7A.4, the one case where the bias of the indirectly standardized relative risk can be eliminated is given.

7A.3.1   Relative Risk.

We will show that when the relative risk is constant and equal to θ within each category of the confounding factor category, direct standardization will be unbiased in large samples. On the other hand, the indirectly standardized relative risk can remain biased, no matter how large the samples.*

Consider a two-category risk factor—present ($r = 1$) and absent ($r = 0$)—and an arbitrary standard population. From Section 7A.2 the directly standardized relative risk is

$$\hat{\theta}^{\mathrm{D}} = \frac{p_1^{\mathrm{D}}}{p_0^{\mathrm{D}}} = \frac{\sum_{k=1}^{K} N_k p_{1k}}{\sum_{k=1}^{K} N_k p_{0k}}, \tag{7.2}$$

and the indirectly standardized relative risk is

$$\hat{\theta}^{\mathrm{I}} = \frac{p_1^{\mathrm{I}}}{p_0^{\mathrm{I}}} = \frac{\mathrm{SMR}_1}{\mathrm{SMR}_0}$$

$$= \frac{\sum_{k=1}^{K} n_{1k} p_{1k} \Big/ \sum_{k=1}^{K} n_{1k} P_k}{\sum_{k=1}^{K} n_{0k} p_{0k} \Big/ \sum_{k=1}^{K} n_{0k} P_k}. \tag{7.3}$$

---

* To be precise, direct standardization yields a consistent estimate of θ; indirect standardization does not.

Suppose that the sample relative risks within each category of the confounding factor are all equal to θ, that is, $p_{1k} = \theta p_{0k}$ for all k. (This will be approximately the case in large samples within each stratum.) Then we have, from (7.2),

$$\hat{\theta}^D = \frac{\sum_{k=1}^{K} N_k \theta p_{0k}}{\sum_{k=1}^{K} N_k p_{0k}} = \theta,$$

regardless of the choice of standard population. Direct standardization is doing the right thing by yielding the common value, θ, as the standardized relative risk. For the indirectly standardized relative risk, on the other hand, we have from (7.3):

$$\hat{\theta}^I = \theta \left(\frac{\sum_{k=1}^{K} n_{1k} p_{0k} \Big/ \sum_{k=1}^{K} n_{1k} P_k}{\sum_{k=1}^{K} n_{0k} p_{0k} \Big/ \sum_{k=1}^{K} n_{0k} P_k} \right),$$

which is not, in general, equal to θ. If, instead, we take the standard population to be one of the risk factor groups, say $r = 0$, so that $P_k = p_{0k}$ for all k, we have

$$\hat{\theta}^I = \frac{\sum_{k=1}^{K} n_{1k} p_{1k}}{\sum_{k=1}^{K} n_{1k} p_{0k}}$$

$$= \frac{\sum_{k=1}^{K} n_{1k} p_{0k} \theta}{\sum_{k=1}^{K} n_{1k} p_{0k}} = \theta.$$

The result of this section, taken together with the result to be presented in Section 7A.4, says that there is only one case where indirect standardization does the right thing in terms of properly estimating the relative risk, but in that case the same answer can be obtained by direct standardization. From a bias point of view, there is thus no reason for choosing indirect standardization.

7A.3.2 Difference of Rates. Now consider direct and indirect standardization as estimators of the difference of rates. For direct standardization,

$$p_1^D - p_0^D = \frac{1}{N_.} \sum_{k=1}^{K} N_k (p_{1k} - p_{0k}).$$

If the expected value of $p_{1k} - p_{0k}$ is some constant Δ for each category of the confounding factor, then the difference of the directly standardized rates is an unbiased estimator of Δ for any sample size.

For indirect standardization, the difference of standardized rates is

$$p_1^I - p_0^I = P(\mathrm{SMR}_1 - \mathrm{SMR}_0)$$

$$= P \left(\frac{\sum_{k=1}^{K} n_{1k}p_{1k}}{\sum_{k=1}^{K} n_{1k}P_k} - \frac{\sum_{k=1}^{K} n_{0k}p_{0k}}{\sum_{k=1}^{K} n_{0k}P_k} \right). \tag{7.4}$$

The expectation of this difference will be something other than Δ regardless of the sample size, except for one very special case.

Take the standard to be the risk-factor-absent group, as in Section 7A.3.1. Although the difference of indirectly standardized rates is still biased, the form of the estimator is informative. Substituting p_{0k} for P_k and p_0^C for P in (7.4), we obtain

$$p_1^I - p_0^I = p_0^C \left[\frac{\sum_{k=1}^{K} n_{1k}(p_{1k} - p_{0k})}{\sum_{k=1}^{K} n_{1k}p_{0k}} \right].$$

Now suppose that, for all k, $p_{1k} - p_{0k} = \Delta$, as would be approximately the case in large samples. Then,

$$p_1^I - p_0^I = \Delta \left[\frac{(1/n_{0.}) \sum_{k=1}^{K} n_{0k}p_{0k}}{(1/n_{1.}) \sum_{k=1}^{K} n_{1k}p_{0k}} \right]. \tag{7.5}$$

This means that, in large samples, $p_1^I - p_0^I$ will be biased unless $\Delta = 0$, and that the greater the confounding, the greater the bias. [The term in brackets in (7.5) can be viewed as a measure of the extent to which the confounding factor distribution in the two risk factor groups differ.]

7A.4   Equivalence of Direct and Indirect Standardization

Whether or not the relative risk is constant within each category of the confounding factor, as was assumed in Section 7A.3, direct and indirect standardization are equivalent if the standard population is taken to be one of the risk factor groups. First, equivalent means that, by choosing the standard appropriately for each type of standardization, the two methods will yield the same estimate of the relative risk.

Suppose that the risk-factor-present group ($r = 1$) is chosen as the standard for direct standardization. Then,

$$N_k = n_{1k} \qquad \text{for all } k,$$

and consequently,

$$p_1^D = p_1^C.$$

For the risk-factor-absent group,

$$p_0^D = \frac{1}{n_1.} \sum_{k=1}^{K} n_{1k} p_{0k},$$

and therefore

$$\hat{\theta}^D = \frac{\sum_{k=1}^{K} n_{1k} p_{1k}}{\sum_{k=1}^{K} n_{1k} p_{0k}}.$$

If the other risk factor group, the risk-factor-absent group ($r = 0$), is chosen as the standard for indirect standardization, then

$$P_k = p_{0k} \qquad \text{for all } k$$

and

$$p_0^1 = p_0^C.$$

For the risk-factor-present group,

$$p_1^1 = p_0^C \frac{\sum_{k=1}^{K} n_{1k} p_{1k}}{\sum_{k=1}^{K} n_{1k} p_{0k}}$$

and therefore

$$\hat{\theta}^I = \frac{\sum_{k=1}^{K} n_{1k} p_{1k}}{\sum_{k=1}^{K} n_{1k} p_{0k}} = \hat{\theta}^D.$$

Note, however, that neither the two sets of standardized rates nor their differences are equal; that is, $p_0^1 \neq p_0^D$, $p_1^1 \neq p_1^D$, and $p_1^1 - p_0^1 \neq p_1^D - p_0^D$.

APPENDIX 7B   STRATIFIED ESTIMATORS OF THE ODDS RATIO

In this appendix various mathematical details regarding the five principal estimators—maximum likelihood, conditional maximum likelihood, Woolf, modified Woolf, and Mantel–Haenszel—of the odds ratio will be given. The estimators due to Birch and Goodman will not be considered, since they are not approximately unbiased in large samples.

The notation for sample quantities is given in Section 7A.1. In addition, the population rate for the kth confounding factor category and rth risk factor category is denoted P_{rk} ($r = 0, 1$ and $k = 1, \ldots, K$). The no-interaction assumption is that the odds ratio is the same in each confounding factor stratum:

$$\psi = \frac{P_{1k}Q_{0k}}{Q_{1k}P_{0k}} \qquad \text{for } k = 1, \cdots, K, \tag{7.6}$$

where $Q_{rk} = 1 - P_{rk}$.

7B.1   Maximum Likelihood and Conditional Likelihood Estimators

The likelihood (ignoring the binomial coefficients) is

$$L = \prod_{k=1}^{K} P_{1k}^{s_{1k}} Q_{1k}^{n_{1k} - s_{1k}} P_{0k}^{s_{0k}} Q_{0k}^{n_{0k} - s_{0k}},$$

where the s_{rk} are the numbers of "successes" ($p_{rk} = s_{rk}/n_{rk}$). In this form there appear to be $2K$ parameters to estimate, the P_{rk}, but actually there are only $K + 1$ independent parameters, owing to the no-interaction assumption (7.6). To reparametrize, let γ denote the natural log of the odds ratio ψ and let

$$\rho_k = \ln\left(\frac{P_{1k}}{Q_{1k}}\right) \qquad \text{for } k = 1, \cdots, K.$$

The natural log of the likelihood is then

$$l = \sum_{k=1}^{K} [s_{1k}\rho_k + n_{1k} \ln Q_{1k} + s_{0k}(\rho_k - \gamma) + n_{0k} \ln Q_{0k}], \tag{7.7}$$

where the Q_{rk} are functions of γ and the ρ_k.

The (usual) maximum likelihood estimator of γ is found by differentiating (7.7) with respect to γ and the ρ_k and then solving for the $K + 1$ unknowns. If the sample sizes within each stratum, the n_{rk}, are all large, the maximum likelihood estimator of γ, $\hat{\gamma}_{\text{ML}}$, will be approximately normally distributed with mean γ and variance W^{-1} (Gart, 1962), where

$$W = \sum_{k=1}^{K} [(n_{0k}P_{0k}Q_{0k})^{-1} + (n_{1k}P_{1k}Q_{1k})^{-1}]^{-1}. \tag{7.8}$$

The maximum likelihood estimator of the odds ratio ψ is

$$\hat{\psi}_{ML} = \exp(\hat{\gamma}_{ML}),$$

which will be approximately normally distributed with mean ψ and variance ψ^2/W. The variance of $\hat{\psi}_{ML}$ can be estimated by replacing each P_{rk} in (7.8) with the corresponding p_{rk}.

This maximum likelihood estimator is identical to that obtained by logit analysis (Chapter 9), using the method of Section 9.8 to handle a confounding factor with more than two categories.

In the alternative asymptotic case, where the number of categories, K, increases, the maximum likelihood estimator given above is not appropriate; as the number of categories increases, the number of parameters also increases, violating one of the assumptions required for the properties of maximum likelihood estimators to hold. In such cases, an alternative maximum likelihood estimator, the conditional maximum likelihood estimator, is appropriate. The term "conditional" comes from the fact that this procedure is based on the likelihood of ψ conditioned on the sufficient statistics for the K nuisance parameters, the ρ_k. This likelihood is (Gart, 1970)

$$L' = \prod_{k=1}^{K} \frac{\binom{n_{1k}}{s_{1k}}\binom{n_{0k}}{t_k - s_{1k}} \psi^{s_{1k}}}{\sum_{j=\max(0, t_k - n_{0k})}^{\min(t_k, n_{1k})} \binom{n_{1k}}{j}\binom{n_{0k}}{t_k - j} \psi^{j}}, \tag{7.9}$$

where $t_k = s_{1k} + s_{0k}$ is the sufficient statistic for ρ_k. The conditional maximum likelihood estimator of ψ, $\hat{\psi}_{CML}$, is that value of ψ which maximizes L'. Thomas (1975) considers the numerical problem of solving for the maximizing value.

Andersen (1970) considers the properties of conditional maximum likelihood estimators in general. Applying his results to this problem, we obtain that $\hat{\psi}_{CML}$ will be approximately normally distributed with mean ψ when K is large. The variance formula is not illuminating; the variance can be estimated as the reciprocal of the second derivative of $-L'$.

Birch (1964) showed that $\hat{\psi}_{CML}$ is the solution of a polynomial equation that involves an expectation taken with respect to the conditional distributions in (7.9). Gart's (1970) approximation to the conditional maximum likelihood estimator is based on approximating this expected value by using a large sample (large n_{rk}) approximation. This estimator, $\hat{\psi}_{AML}$ (A for asymptotic), is the solution to

$$s_1 = \sum_{k=1}^{K} \hat{s}_k$$

where $s_1 = \sum^K_{k=1} s_{1k}$ and each $\hat{s}_k$ satisfies

$$\frac{\hat{s}_k(n_{0k} - t_k + \hat{s}_k)}{(t_k - \hat{s}_k)(n_{1k} - \hat{s}_k)} = \hat{\psi}_{\mathrm{AML}}.$$

Again, Thomas (1975) considers the numerical problems of solving for $\hat{\psi}_{\mathrm{AML}}$.

Gart's approximation to the conditional maximum likelihood estimator does require large n_{rk} to be valid, but, unlike the approximations due to Birch and Goodman, is valid for all value of the odds ratio, not just $\psi = 1$.

7B.2   Woolf and Modified Woolf Estimators

The estimator of the log odds ratio proposed by Woolf (1955) is a weighted average of the estimators of the log odds ratio from each of the strata:

$$\hat{\gamma}_w = \frac{\sum\limits_{k=1}^{K} w_k \hat{\gamma}_k}{w}$$

where

$$\hat{\gamma}_k = \ln\left(\frac{p_{1k}q_{0k}}{q_{1k}p_{0k}}\right) = \ln\left[\frac{s_{1k}(n_{0k} - s_{0k})}{(n_{1k} - s_{1k})s_{0k}}\right] \tag{7.10}$$

$$w_k = [(n_{0k}p_{0k}q_{0k})^{-1} + (n_{1k}p_{1k}q_{1k})^{-1}]^{-1}$$

$$= \left[\frac{n_{0k}}{s_{0k}(n_{0k} - s_{0k})} + \frac{n_{1k}}{s_{1k}(n_{1k} - s_{1k})}\right]^{-1}$$

$$w = \sum_{k=1}^{K} w_k. \tag{7.11}$$

w_k^{-1} is an estimate of the variance of $\hat{\gamma}_k$, so the Woolf estimator is based on weighting inversely proportional to the variance.

From (7.10) it is clear that the Woolf estimator cannot be calculated if any p_{rk} or q_{rk} is zero. A modification that avoids this problem is based on work of Haldane (1955) and Anscombe (1956). They independently showed that a less biased estimator of the log odds ratio from the kth confounding factor stratum is

$$\hat{\gamma}'_k = \ln\left[\frac{(s_{1k} + 0.5)(n_{0k} - s_{0k} + 0.5)}{(n_{1k} - s_{1k} + 0.5)(s_{0k} + 0.5)}\right]; \tag{7.12}$$

that is, just add 0.5 to each of the four quantities in the sample odds ratio formula. Gart (1966) suggested a modified Woolf estimator of the form

$$\hat{\gamma}_{MW} = \frac{\sum\limits_{k=1}^{K} w'_k \hat{\gamma}'_k}{w'},$$

where

$$w'_k = \left[\frac{n_{0k} + 1}{(s_{0k} + 0.5)(n_{0k} - s_{0k} + 0.5)} + \frac{n_{1k} + 1}{(s_{1k} + 0.5)(n_{1k} - s_{1k} + 0.5)}\right]^{-1}$$

$$w' = \sum_{k=1}^{K} w'_k. \tag{7.13}$$

The results of Gart and Zweifel (1967), who considered various estimators of the log odds and estimators of the variances of the log odds estimators, suggest that $(w'_k)^{-1}$ is generally the least biased estimator of the variance of $\hat{\gamma}'_k$. [An alternative modification to the weights, not considered here, was suggested by Haldane (1955).]

For the asymptotic case of large n_{rk}, the Woolf and modified Woolf estimators have the same large-sample distribution as the maximum likelihood estimator. In particular, the asymptotic variances are equal, so the two Woolf estimators, as well as the maximum likelihood estimator, are asymptotically efficient. Estimated variances for the Woolf and modified Woolf estimators are w^{-1} (7.11) and $(w')^{-1}$ (7.13), respectively.

7B.3   Mantel–Haenszel Estimator

Mantel and Haenszel (1959) proposed the estimator

$$\hat{\psi}_{MH} = \sum_{k=1}^{K} \frac{s_{1k}(n_{0k} - s_{0k})}{n_{1k} + n_{0k}} \bigg/ \sum_{k=1}^{K} \frac{(n_{1k} - s_{1k})s_{0k}}{n_{1k} + n_{0k}}$$

$$= \sum_{k=1}^{K} \frac{m_k \hat{\psi}_k}{m},$$

where

$$m_k = \left(\frac{1}{n_{1k}} + \frac{1}{n_{0k}}\right)^{-1} q_{1k} p_{0k}$$

$$m = \sum_{k=1}^{K} m_k.$$

Hauck (1979) has shown that if the n_{rk} are large, the Mantel–Haenszel estimator is approximately normally distributed with mean ψ and variance

$$V = \frac{\psi^2 \sum\limits_{k=1}^{K} M_k^2 W_k^{-1}}{M^2}, \tag{7.14}$$

where

$$M_k = \left(\frac{1}{n_{1k}} + \frac{1}{n_{0k}}\right) Q_{1k}P_{0k}$$

$$M = \sum_{k=1}^{K} M_k$$

$$W_k^{-1} = (n_{0k}P_{0k}Q_{0k})^{-1} + (n_{1k}P_{1k}Q_{1k})^{-1}.$$

A sufficient condition for the variance of the Mantel–Haenszel estimator (7.14) to be equal to that of the maximum likelihood estimator (ψ^2/W) is $\psi = 1$.

REFERENCES

Andersen, E. B. (1970), Asymptotic Properties of Conditional Maximum Likelihood Estimators, *Journal of the Royal Statistical Society, Series B*, **32**, 283–301.

Anscombe, F. J. (1956), On Estimating Binomial Response Relations, *Biometrika*, **43**, 461–464.

Birch, M. M. (1964), The Detection of Partial Association I: The 2×2 Case, *Journal of the Royal Statistical Society, Series B*, **26**, 313–324.

Bishop, Y. M. M. (1967), Multidimensional Contingency Tables: Cell Estimates, Ph.D. thesis, Harvard University.

Bishop, Y. M. M., Fienberg, S. E., and Holland, P. W. (1974), *Discrete Multivariate Analysis: Theory and Practice*, Cambridge, MA: MIT Press.

Chiang, C. L. (1961), Standard Error of the Age-Adjusted Death Rate, U.S. Dept. Health, Education and Welfare, Vital Statistics, Special Report 47, pp. 271–285.

Cochran, W. G. (1954), Some Methods for Strengthening the Common χ^2 Tests, *Biometrics*, **10**, 417–451.

Cochran, W. G. (1968), The Effectiveness of Adjustment by Subclassification in Removing Bias in Observational Studies, *Biometrics*, **24**, 295–313.

Fleiss, J. L. (1973), *Statistical Methods for Rates and Proportions*, New York: Wiley.

Gart, J. J. (1962), On the Combination of Relative Risks, *Biometrics*, **18**, 601–610.

Gart, J. J. (1966), Alternative Analyses of Contingency Tables, *Journal of the Royal Statistical Society, Series B*, **28**, 164–179.

Gart, J. J. (1970), Point and Interval Estimation of the Common Odds Ratio in the Combination of 2×2 Tables with Fixed Marginals, *Biometrika*, **57**, 471–475.

Gart, J. J. (1971), The Comparison of Proportions: A Review of Significance Tests, Confidence Intervals and Adjustments for Stratification, *Review of the International Statistical Institute*, **39**, 148–169.

Gart, J. J., and Zweifel, R. (1967), On the Bias of Various Estimators of the Logit and Its Variance with Application to Quantal Bioassay, *Biometrika*, **54**, 181–187.

Goldman, A. (1971), The Comparison of Multidimensional Rate Tables—A Simulation Study, Ph.D. thesis, Harvard University.

Goodman, L. A. (1969), On Partitioning χ^2 and Detecting Partial Association in Three-Way Tables, *Journal of the Royal Statistical Society, Series B*, **31**, 486–498.

Haldane, J. B. S. (1955), The Estimation and Significance of the Logarithm of a Ratio of Frequencies, *Annals of Human Genetics,* **20,** 309–311.

Hauck, W. W. (1979), The Large Sample Variance of the Mantel–Haenszel Estimator of a Common Odds Ratio, *Biometrics,* **35,** 817–819.

Herring, R. A. (1936), The Relationship of Martial Status in Females to Mortality from Cancer of the Breast, Female Genital Organs and Other Sites, *The American Society for the Control of Cancer,* **18,** 4–8.

Kalton, G. (1968), Standardization: A Technique to Control for Extraneous Variables, *Applied Statistics,* **17,** 118–136.

Keyfitz, N. (1966), Sampling Variance of Standardized Mortality Rates, *Human Biology,* **3,** 309–317.

Kitagawa, E. M. (1964), Standardized Comparisons in Population Research, *Demography,* **1,** 296–315.

Kitagawa, E. M. (1966), Theoretical Considerations in the Selection of a Mortality Index, and Some Empirical Comparisons, *Human Biology,* **38,** 293–308.

McKinlay, S. M. (1975), The Effect of Bias on Estimators of Relative Risk for Pair Matched and Stratified Samples, *Journal of the American Statistical Association,* **70,** 859–864.

McKinlay, S. M. (1978), The Effect of Nonzero Second-Order Interaction on Combined Estimators of the Odds Ratio, *Biometrika,* **65,** 191–202.

Mantel, N., and Haenszel, W. (1959), Statistical Aspects of the Analysis of Data from Retrospective Studies of Disease, *Journal of the National Cancer Institute,* **22,** 719–748.

Miettinen, O. S. (1972), Standardization of Risk Ratios, *American Journal of Epidemiology,* **96.** 383–388.

Miettinen, O. S. (1976), Stratification by a Multivariate Confounder Score, *American Journal of Epidemiology,* **104,** 609–620.

Spiegelman, M., and Marks, H. H. (1966), Empirical Testing of Standards for the Age Adjustment of Death Rates by the Direct Method, *Human Biology,* **38,** 280–292.

Thomas, D. G. (1975), Exact and Asymptotic Methods for the Combination of 2 × 2 Tables, *Computers and Biomedical Research,* **8,** 423–446.

Woolf, B. (1955), On Estimating the Relation between Blood Group and Disease, *Annals of Human Genetics,* **19,** 251–253.

CHAPTER 8

Analysis of
Covariance

In this chapter we consider an adjustment strategy that is appropriate for cohort studies with a numerical outcome factor, a categorical treatment (or risk) factor, and a numerical confounding factor. Under these conditions, the *general linear model* can be applied to the problem of estimating treatment effects. The *analysis of covariance* (ANCOVA) represents the main application of the linear model for this purpose.

8.1 BACKGROUND

The general linear model represents the outcome value as a linear combination (weighted sum) of measured variables. Generally speaking, when these variables are all numerical, the linear model is called a *regression model*. When the variables are all categorical, we refer to the *analysis of variance* (ANOVA).

While both regression and analysis of variance can be formally subsumed under the general linear model, the two techniques have traditionally been treated as distinct. This historical separation occurred for two reasons. First, before high-speed computers were in general use, computational aspects of statistical techniques were of much interest. The most efficient computational procedures for regression and ANOVA were quite different. Second, the two methods tended to be applied to different sorts of problems.

The analysis of variance is usually thought of as a technique for comparing the means of two or more populations on the basis of samples from each. In practice, these populations often correspond to different treatment groups, so that differences in population means may be evidence for corresponding differences in treatment effects.

The ANOVA calculations involve a division of the total sample variance into within-group and between-group components. The within-group component provides an estimate of error variance, while the between-group component estimates error variance plus a function of the differences among treatment means. The ratio of between- to within-group variance provides a test of the null hypothesis that all means are equal. Moreover, the differences among group means provide unbiased estimates of the corresponding population mean differences, and standard errors based on the within-group variance provide confidence intervals for these differences and tests of their significance.

Regression analysis, on the other hand, is primarily used to model relationships between variables. With it, we can estimate the form of a relationship between a response variable and a number of inputs. We can try to find that combination of variables which is most strongly related to the variation in the response.

The analysis of covariance represents a marriage of these two techniques. Its first use in the literature was by R. A. Fisher (1932), who viewed the technique as one that "combines the advantages and reconciles the requirements of the two very widely applicable procedures known as regression and analysis of variance."

Combining regression and ANOVA provides the powerful advantage of making possible comparisons among treatment groups differing prior to treatment. Suppose we can identify a variable X that is related to the outcome, Y, and on which treatment groups have different means. We shall assume for simplicity that X is the only variable on which the groups differ. Then, if we knew the relationship between Y and X, we could appropriately adjust the observed differences on Y to take account of the differences on X.

8.2 EXAMPLE: NUTRITION STUDY COMPARING URBAN AND RURAL CHILDREN

Greenberg (1953) described a nutrition study designed to compare growth of children in an urban environment with that of rural children. Data were obtained on the heights of children in the two samples: one from an urban private school and one from a rural public school. Differences in growth between these groups might be the result of the different environmental influences operating on the children. In particular, the rural children might be experiencing some nutritional deprivation relative to their urban counterparts. In the terminology of this book, height would be the response or outcome factor and nutrition the risk factor of interest.

The data are shown in Table 8.1. An analysis of variance conducted on the height data reveals that the observed difference between the groups (2.8 cm) is not statistically significant. So it might be concluded that there is no evidence here for a difference in nourishment between the urban and rural school children.

Table 8.1 Height and Age of Private and Rural School Children in a Study in North Carolina in 1948

| | Private School | | Rural School | |
| | Age (months) | Height (cm) | Age (months) | Height (cm) |
Students				
1	109	137.6	121	139.0
2	113	147.8	121	140.9
3	115	136.8	128	134.9
4	116	140.7	129	149.5
5	119	132.7	131	148.7
6	120	145.4	132	131.0
7	121	135.0	133	142.3
8	124	133.0	134	139.9
9	126	148.5	138	142.9
10	129	148.3	138	147.7
11	130	147.5	138	147.7
12	133	148.8	140	134.6
13	134	133.2	140	135.8
14	135	148.7	140	148.5
15	137	152.0		
16	139	150.6		
17	141	165.3		
18	142	149.9		
Mean	126.8	144.5	133.1	141.7

Reprinted, by permission, from Greenberg (1953), Table 1.

Before reaching this conclusion, however, we should consider whether there are likely to be confounding factors. One variable that comes immediately to mind is age. The data on age are also presented in Table 8.1. The mean age for the rural children is 6.3 months greater than that of the urban children. In a sense, then, the rural children have an "unfair advantage" conferred by their greater average age. Thus we might expect that if the age distributions were the same, the difference in average height between the groups would be even larger than the observed 2.8 cm. The analysis of covariance allows us to adjust the 2.8-cm difference to obtain a better (less-biased) estimate of the difference between groups that *would* have been observed had the mean ages in the two groups been equal. As we shall see in Section 8.3, ANCOVA produces an estimated difference of 5.5 cm, which is significant at the .05 level.

In addition to the bias reduction described above, another benefit results from the combination of regression analysis and ANOVA. Suppose that *within* treatment groups, a substantial proportion of the variance in Y can be explained by variation in X. In carrying out an ANOVA, we would like the within-group variance to reflect only random error. Regression analysis can be used to remove that part of the error attributable to X and thereby to increase the precision of group comparisons.

The Greenberg (1953) example mentioned above can be used to illustrate this point as well. It is clear from Table 8.1 that a substantial proportion of the variation in height is attributable to variation in age. Put differently, if all children in a group were of the same age, the variation in heights within that group would be substantially reduced. Since the relationship between height and age over this range is quite linear, we can estimate the pure error variation by taking residuals* around the regression line relating the two variables. In effect, this is what ANCOVA does, and when a high proportion of within-group variance is explained by the covariate, a large increase in precision results.

In summary, then, ANCOVA combines the advantages of regression and ANOVA in comparing treatments by providing two important benefits. First, by estimating the form of the relationship between outcome and covariate, an appropriate adjustment can be made to remove biases resulting from group differences on the covariate. This advantage is of importance primarily in nonrandomized studies, where such group differences are likely to occur. Second, by reducing the variation within groups, the precision of estimates and tests used to compare groups can be increased. This advantage may be valuable in both randomized and nonrandomized studies.

By combining the advantages of ANOVA and regression, ANCOVA provides a powerful tool for estimating treatment effects. As noted by Fisher, however, the technique also "reconciles the requirements" of the techniques. Thus, to be valid, the ANCOVA must be used in situations satisfying the requirements for

---

* The *residual* corresponding to a given observation is defined as the difference between the actual observed Y and the value predicted by substituting the corresponding X value into the regression equation.

both techniques. Put differently, the usefulness of the ANCOVA rests on the validity of a certain mathematical model for the generation of data, which in turn rests on a set of assumptions. To obtain the advantages of both regression and ANOVA, we must be willing to assert that a somewhat restrictive model is valid.

In the remainder of this chapter, we will attempt to provide enough understanding of the rationale and assumptions underlying ANCOVA to enable the reader to understand when ANCOVA can be used and how to interpret the results generated. Since the actual calculations involved in carrying out the analysis are complex, they are almost always performed by a computer, and it would be unnecessarily confusing to present the formulas here. For the reader interested in more detail, a technical appendix containing some basic formulas is included at the end of this chapter. More extensive discussions can be found in Cochran (1957) and Winer (1971, Chap. 10).

8.3   THE GENERAL ANCOVA MODEL AND METHOD

To understand the rationale underlying the use of ANCOVA in nonrandomized studies, it is helpful to begin with a somewhat idealized situation. Suppose that on the basis of extensive prior research, the relationship between an outcome and confounding factor can be specified. For example, it might be known that for rural school children, the relationship between height and age over the age range being considered can be expressed as

$$\text{Average height} = 75 + 0.5 \,(\text{age}).$$

Suppose that a particular group of rural children have been exposed to some special treatment, such as a dietary supplement. At the time they are measured this group has a mean age of 132 months and a mean height of 147 cm. Suppose further that another group has been exposed to a different treatment and is measured when the children are 120 months old on the average. The average height of this group is 133 cm.

Since the groups differ on mean age, it is not obvious which treatment has been more effective. To make a fair comparison, we must remove the effect of the confounding variable age. However, using the relationship specified above we know that the expected height for the two groups without any special treatment is given by:

Group 1:   Average height = 75 + 0.5(132) = 141 cm
Group 2:   Average height = 75 + 0.5(120) = 135 cm.

Therefore, the effects of the treatments are:

Group 1:   Effect = observed − expected = 147 − 141 = 6 cm
Group 2:   Effect = observed − expected = 133 − 131 = −2 cm.

and the difference between them is 8 cm.

Alternatively, we can say that because the groups differ by 12 months in age, the relationship predicts that they will differ by 6 cm. So we could effectively "adjust" the comparison between the two groups by subtracting 6 cm from the difference between them. Since the observed difference is 14 cm, this would leave 8 cm attributable to the difference in treatments received..

Because we are assuming in this example a known baseline relationship against which to measure performance under the treatments, we can obtain an absolute measure of effect for each treatment (6 cm and −2 cm). In most practical situations, we do not have available such an external standard, and we must use only data obtained during the study. Thus an absolute measure of effect for each group is impossible. On the other hand, it may still be possible to obtain from the data an estimate of the coefficient (0.5 cm/month in our example) relating outcome level to confounding variable. So it may be possible to adjust the observed difference to remove the effect of age from the comparison. In effect, this is how ANCOVA is used to estimate treatment effects in nonrandomized comparative studies.

The basic model underlying the use of the standard analysis of covariance asserts that there is a linear relationship between the outcome Y and the covariate X with identical slopes in the two groups, but possibly different intercepts. With two treatment groups, we can write the basic model as*

$$Y = \alpha_1 + \beta X + e \quad \text{in group 1 (treatment)}$$
$$Y = \alpha_0 + \beta X + e \quad \text{in group 0 (control),}$$

$$(8.1)$$

where

α_1 = expected value of Y when $X = 0$ for group 1

α_0 = expected value of Y when $X = 0$ for group 0

e = random variable representing error
(expectation 0 for any given X).

Let $\overline{X}$ represent the sample mean of all the X observations in both groups, $\overline{X}_1$, the mean for group 1, and $\overline{X}_0$ the mean for group 0. Figure 8.1 illustrates this situation. Note that the direct comparison of $\overline{Y}_1$ and $\overline{Y}_0$ will be biased since $\overline{X}_1$

---

* For the reader familiar with regression analysis, this model can be represented as a two-variable regression model with variables X and a dummy variable taking the value 1 in group 1 and 0 in group 0.

$\neq \overline{X}_0$. In fact, taking means in (8.1) yields

$$\overline{Y}_1 = \alpha_1 + \beta\overline{X}_1 + \overline{e}_1$$
$$\overline{Y}_0 = \alpha_0 + \beta\overline{X}_0 + \overline{e}_0,$$

so that

$$E(\overline{Y}_1 - \overline{Y}_0) = \alpha_1 - \alpha_0 + \beta(\overline{X}_1 - \overline{X}_0). \tag{8.2}$$

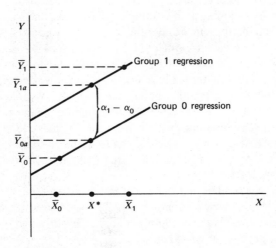

Figure 8.1.   Standard ANCOVA assumptions.

Note that from (8.1), we can interpret $\alpha_1 - \alpha_0$ as the expected difference between the outcomes of the two individuals with the same value of X but in two different groups. This difference will represent the differential effect of the two treatments unless there is some other variable related to Y which distinguishes the two subjects. To estimate $\alpha_1 - \alpha_0$, we cannot simply subtract $\overline{Y}_0$ from $\overline{Y}_1$, but must adjust each of these to move them, in effect, to a common X value, say X^*. Let us define the "adjusted" mean of Y for group 1 as

$$\overline{Y}_{1a} = \overline{Y}_1 - \beta(\overline{X}_1 - X^*).$$

$\overline{Y}_{1a}$ may be interpreted as an estimate of the mean outcome for members of group 1 whose X value is X^*. Similarly,

$$\overline{Y}_{0a} = \overline{Y}_0 - \beta(\overline{X}_0 - X^*)$$

estimates the mean outcome for members of group 0 whose X value is X^*. To estimate the difference between the means of the two groups at the same value of X (in this case X^*), we can simply take the difference of these two adjusted means:

$$\overline{Y}_{1a} - \overline{Y}_{0a} = \overline{Y}_1 - \beta(\overline{X}_1 - X^*) - [\overline{Y}_0 - \beta(\overline{X}_0 - X^*)]$$
$$= \overline{Y}_1 - \overline{Y}_0 - \beta(\overline{X}_1 - \overline{X}_0). \tag{8.3}$$

This adjusted group mean difference is an unbiased estimator of $\alpha_1 - \alpha_0$.

For simplicity, we have not discussed how the value of β necessary to perform the adjustments is actually obtained. In practice, we rarely have any a priori theoretical basis for determining the value of β and must therefore use the data to obtain an estimate, $\hat{\beta}$. The ANCOVA calculations provide us with an unbiased estimator based on the relationship between Y and X within the two groups. Thus the adjusted difference is of the form

$$\overline{Y}_{1a} - \overline{Y}_{0a} = \overline{Y}_1 - \overline{Y}_0 - \hat{\beta}(\overline{X}_1 - \overline{X}_0). \tag{8.4}$$

It can be shown that the substitution of an unbiased estimate $\hat{\beta}$ for the unknown true value β still yields an unbiased estimate of $\alpha_1 - \alpha_0$ under the model specified by (8.1).

In Appendix 8A we present the formula usually used to compute $\hat{\beta}$. It is called a pooled within-group regression coefficient, because it combines data on the relationship between Y and X in both groups. This combination of data provides high precision and is valid under our assumption that the regression lines are parallel.

We should mention in passing that this pooled coefficient is not found by calculating a regression coefficient from the data on both groups taken together as a single group, as is sometimes proposed. This latter approach may be viewed as comparing the mean residuals for the two groups around the *overall* regression line fitted to the entire sample. It is incorrect, however, in the sense that it does not yield an unbiased estimate of β or of the effect $\alpha_1 - \alpha_0$ under the model given by (8.1).

Using the standard ANCOVA calculations (see Appendix 8A), we obtain for the Greenberg (1953) example:

$$\hat{\beta} = 0.42 \text{ cm/month},$$

and because

$$\overline{X}_1 = 126.8 \text{ months}$$

and

$$\overline{X}_0 = 133.1 \text{ months},$$

the adjusted difference is

$$\overline{Y}_{1a} - \overline{Y}_{0a} = 2.8 - 0.42(126.8 - 133.1) = 5.5 \text{ cm}.$$

The initial difference of 2.8 cm in favor of the urban children has, after adjustment, been nearly doubled.

We may ask at this point whether this adjusted difference is statistically significant. To answer this question, we can look at the standard error provided as part of the ANCOVA calculations. This standard error can be used to perform a t test of

$$H_0 : \alpha_1 = \alpha_0.$$

More generally, when there are more than two treatment groups (say K groups), ANCOVA provides an F test of

$$H_0 = \alpha_1 = \alpha_2 = \alpha_3 = \cdots = \alpha_K.$$

If this test proves significant, we can reject the null hypothesis that all treatment groups have the same intercept. In this case we must conclude either that the treatments are differentially effective or that there is some unmeasured variable related to outcomes on which the groups vary (i.e., another confounding factor). In the Greenberg (1953) example, a t test for the difference of adjusted means results in a t value of 2.12, which is significant at the .05 level. So when age is taken into account, there appears to be a significant difference in height between the two samples.

8.4 ASSUMPTIONS UNDERLYING THE USE OF ANCOVA

In Section 8.3 we presented the basic model underlying the use of ANCOVA in the simple situation with two treatment groups and one covariate. This model is summarized by (8.1). While this statement of the model appears simple, it implies a large number of conditions that must be satisfied. Since the user must verify that these conditions hold, we present in this section a listing of the assumptions. With each of these, we indicate the consequences of failures to satisfy the assumption and how these can be detected in practice. The next section considers some ways of reducing the biases introduced by such failures.

Like any mathematical model attempting to represent reality, the ANCOVA model is never perfectly true. It is only a more-or-less accurate abstraction. So, although we may for simplicity discuss whether or not a particular condition holds in a particular situation, it should be remembered that such statements are only approximate. The real question is whether the ANCOVA model is good enough not to result in misleading results. With this caveat in mind, we now proceed to list the ANCOVA assumptions.

1. *Equality of regression slopes.* ANCOVA assumes that the relationship between Y and X in each group differs only in terms of the intercept (α_1) but

not the slope (β). This assumption is essential if we are to have the *possibility* of interpreting the difference between the lines ($\alpha_1 - \alpha_0$) as a measure of treatment effect. The problem of nonparallel regressions in different treatment groups is discussed in Section 3.3 and is a general problem involved in all adjustment strategies. The nature of the difficulty is illustrated in Figure 8.2. The expected difference between two individuals in different groups with identical X values depends on X. Thus there is no unique summary value which can be interpreted as *the* treatment effect.

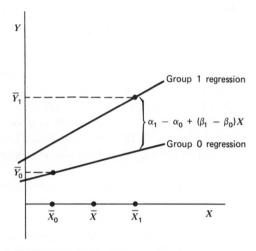

Figure 8.2. Nonparallel linear regressions in two groups.

In such a situation we say there is an *interaction* between the treatment effect and the covariate. If an interaction is suspected, it is worthwhile to examine carefully the graph of Y versus X in the two groups. Visual inspection will usually be adequate to detect serious departures from parallelism.

A formal statistical test for the equality of slopes can also be conducted. If such a test is carried out, and the null hypothesis of slopes rejected, we cannot apply ANCOVA. If, on the other hand, the null hypothesis is not rejected, we still cannot be sure that the slopes are identical. This is a general property of statistical tests. Our ability to assert that the null hypothesis in fact holds if it is not rejected is related to the "power" of the test, which is difficult to compute. Generally speaking, however, the power increases with the sample size. So a statistical test can provide evidence on whether the slopes are equal, but no certainty unless the sample sizes are very large.

2. *Linearity.* The ANCOVA assumes a *linear* relationship between Y and X. The simplest, and usually adequate, test of linearity is to plot a graph of Y

versus X in each group. Formal statistical tests of linearity are available if there is any doubt. The simplest involves calculating the regression line in each group and examining the residuals. Standard texts on regression analysis (e.g., Draper and Smith, 1966; Chatterjee and Price, 1977; Mosteller and Tukey, 1977) provide more detail.

3. *Covariate measured without error.* In some situations, the variable thought to be linearly related to Y cannot be measured directly, and an imperfect substitute containing some measurement error must be used. In Section 5.2 we discussed the issues of measurement error and reliability in some detail. When the observed X, consisting in part of error, is used in the ANCOVA model, both estimates and tests may be affected. In both randomized and nonrandomized studies, the precision of the estimated effect and the power of statistical tests will generally decrease as the reliability decreases. Further, in nonrandomized studies, measurement error will introduce bias in situations where using the true X yields an unbiased estimate (see Cochran, 1968; Elashoff, 1969). When even the true X would result in bias, the effect of measurement error is more complex. Sometimes a fallible variable may even be preferable to a corresponding true score (Weisberg, 1979), although such situations are extremely rare in practice. As a general rule, it is desirable whenever possible to use variables with high reliability.

4. *No unmeasured confounding variables.* The existence of unmeasured variables which are related to the outcome and have unequal distributions in the treatment groups is a general problem in the analysis of nonrandomized studies (Section 5.1). Let us consider what happens when an ANCOVA is performed which does not consider such a variable. Suppose that there exists a variable Z with means $\overline{Z}_1$ and $\overline{Z}_0$ for the groups. Then, instead of (8.1), the true model might be described by

$$Y = \mu = \alpha_i + \beta X + \gamma Z + e \qquad i = 0, 1. \tag{8.5}$$

In this case, the appropriate adjustment becomes

$$\overline{Y}_{ia} = \overline{Y}_i - \beta(\overline{X}_i - \overline{X}) - \gamma(\overline{Z}_i - \overline{Z}) \qquad i = 0, 1.$$

Thus if we adjust using X only as a covariate, and if $\overline{Z}_1 \neq \overline{Z}_0$, we have adjusted for only part of the differences between groups which is related to Y. Further discussion of this issue can be found in Cochran and Rubin (1973), Cronbach et al. (1977), and Weisberg (1979).

5. *Errors independent of each other.* The error terms (e) in the model are random variables which are assumed to be probabilistically independent of one another. This means that the value of the error term corresponding to any observation has no systematic relationship to that of any other error term.

Nonindependence of errors can affect the validity of tests used to compare treatment groups, but will not introduce bias into the estimates of treatment effects. Nonindependence is difficult to detect empirically, and there is usually no reason to suspect its occurrence. However, in some situations there may be theoretical considerations suggesting nonindependence. Suppose, for instance, that the rural children in our example actually came from a small number of families. Then we might expect high correlations between the error terms corresponding to children in the same family. Roughly speaking, the effect of such intercorrelations is to reduce the effective sample size on which inferences are based. That is, the precision is lower than would be expected on the basis of the sample size used.

6. *Equality of error variance.* Ordinarily, as in most applications of linear models, it is assumed that all error terms have the same variance. In an AN-COVA situation, it is possible that the treatment grops have different error variances. The estimates of treatment effects will still be unbiased in this case, but the validity of tests may be affected. If there is some reason to suspect this inequality of error variances, the residuals from the fitted lines in the two groups can be compared. If the variances of these residuals differ greatly, caution in the interpretation of test results is advised (see Glass et al., 1972).

7. *Normality of errors.* For the ANCOVA tests to be strictly valid, it must be assumed that the errors follow a normal distribution. Departures from normality may affect statistical tests and the properties of estimators in a variety of ways, depending on the actual form of the error distribution. The normality assumption can be tested by examining the distribution of residuals. While severe departures from normality may affect the properties of tests, ANCOVA appears to be generally rather robust (see Elashoff, 1969; Glass et al., 1972). Thus most researchers assume that the normality assumption is not critical.

8.5 DEALING WITH DEPARTURES FROM THE ASSUMPTIONS

As indicated in Section 8.4, several assumptions underlie the use of ANCO-VA. Departures from these assumptions may result in biased effect estimates and/or a loss of precision in statistical tests and estimates. While the precision of a statistical procedure is important when the sample size is not large, our primary emphasis in this book has been on the reduction of bias in nonrandomized studies.

In this section we consider what can be done when various departures from the standard ANCOVA assumptions are suspected. Of the seven conditions discussed in Section 8.4, only four bear seriously on the possibility of bias: linearity of the relationship between Y and X, same slopes for regression lines in the two groups, absence of measurement error in the covariate, and absence of other unmeasured covariates.

8.5.1 Nonparallel Regressions

As in Section 8.4, we consider first the case of linear but nonparallel regressions. This is the situation illustrated in Figure 8.2. Since the slopes of the lines, as well as their intercepts, differ in the two groups, the basic model becomes

$$Y + \alpha_i + \beta_i X + e \qquad i = 0, 1. \tag{8.6}$$

From (8.6) the difference between the expected outcomes of the individuals with the same X but in different groups is given by

$$\alpha_1 - \alpha_0 + (\beta_1 - \beta_0)X.$$

That is, the treatment effect is a linear function of X. To estimate this function, we can compute estimates of $\hat{\beta}_1$ and $\hat{\beta}_0$ separately from the two treatment groups and form

$$\overline{Y}_{1a} = \overline{Y}_1 - \hat{\beta}_1(\overline{X}_1 - X)$$
$$\overline{Y}_{0a} = \overline{Y}_0 - \hat{\beta}_0(\overline{X}_0 - X),$$

the treatment means adjusted to an arbitrary point X. Taking the difference yields an unbiased estimated of the treatment effect for any X:

$$\overline{Y}_{1a} - \overline{Y}_{0a} = \overline{Y}_1 - \overline{Y}_0 - \hat{\beta}_1(\overline{X}_1 - X) + \hat{\beta}_0(\overline{X}_0 - X). \tag{8.7}$$

If a single summary value is desired, some "typical" value of X must be inserted in this expression. This might be $\overline{X}$, the mean of X in the two groups together, or the mean from some other standard population. The choice of an X value at which to estimate the treatment effect must be guided by logical rather than statistical considerations. The value should be one that is of practical importance. For example, if we know that the treatment will be applied in the future to individuals with an average value that is at least approximately known, we may whish to estimate the effect at this value.

In many situations, the individuals to receive treatment in the future are expected to be similar to those receiving treatment during the study. So we might wish to choose $X = \overline{X}_1$ in (8.7). We then obtain

$$\overline{Y}_{1a} - \overline{Y}_{0a} = \overline{Y}_1 - \overline{Y}_0 - \hat{\beta}_0(\overline{X}_1 - \overline{X}_0). \tag{8.8}$$

This is of the same form as the standard ANCOVA estimate of the treatment mean difference except that $\hat{\beta}_0$, the estimate based on control group data only has replaced $\hat{\beta}$, the estimate based on pooling the data from both groups. This estimate, first suggested by Belson (1956) and later analyzed by Cochran (1969), is not widely known but offers advantages over the usual estimate in some situations.

Suppose first that the usual ANCOVA model (8.1) holds. In this case the Belson estimate is unbiased but somewhat less precise (larger variance) than the usual estimate. On the other hand, particularly if the control group receives a traditional treatment modality, there may be outside evidence and/or a large sample available to estimate $\hat{\beta}_0$. These factors may outweigh the loss of data from the treatment group.

If the true slopes in the two groups are different, the Belson estimate still has a meaningful interpretation. As noted above, it represents an estimate of the difference in outcomes for individuals with an X value of $\overline{X}_1$. That is, it estimates the effect for a typical individual in the group that received the treatment.

Note that in one sense, (8.7) is more general than the usual ANCOVA model. The usual model represents the special case when $\beta_1 = \beta_0$. On the other hand, unless $\beta_1 = \beta_0$, we cannot use the pooled estimate of β, based on combined data from the two groups. Estimating separate coefficients, as in (8.7), entails the use of smaller samples for each estimated coefficient. For modest sample sizes, this may lead to a slight decrease in precision.

The methods we have so far considered for comparing treatments when regression lines are nonparallel involve specifying a particular covariate value and estimating the effect conditional on this value. If we have some reason for focusing attention on a particular X value, or set of values, this approach will be useful. In some situations, however, we may be more interested in identifying the set of X values for which each treatment is preferable. Figure 8.2 illustrates a situation where for all X values of practical interest, treatment 1 is superior. In Figure 8.3, however, we have a case where treatment 1 is superior for small values of the covariate but inferior for large values. Knowing even approximately where the break-even point is located could have important practical implications.

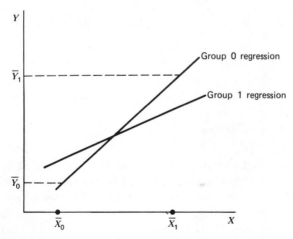

Figure 8.3. Crossing linear regressions in two groups.

Because our estimates of the regression coefficients β_1 and β_0 are subject to sampling variability, we cannot specify the crossing point exactly. However, it is possible to determine a region of X values where the treatment effect is significantly positive or significantly negative, at a specified level of statistical significance. For other values of X, we cannot make a useful statement about which treatment is superior. This approach is known as the Johnson–Neyman technique (Johnson and Neyman, 1936). A good exposition of the technical details and some refinements of the original procedure can be found in Potthoff (1964), and a less technical exposition in Walker and Lev (1959, Chap. 14).

8.5.2 Nonlinear Regressions

The second major threat to the validity of the ANCOVA is nonlinearity of the regressions of Y and X. There are essentially three cases to consider here. The first is illustrated by the solid lines in Figure 8.4: the regressions of Y on X are nonlinear but parallel in the two groups. The treatment effect is in principle clearly defined, but may be difficult to estimate in practice.

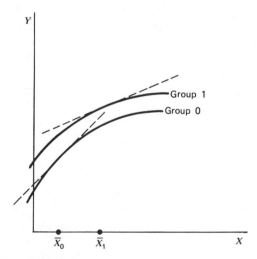

Figure 8.4. Parallel nonlinear regressions.

Let us consider first what happens when we carry out a standard ANCOVA in this situation. Loosely speaking, if $\overline{X}_1$ and $\overline{X}_0$ are not too far apart, and the curvature of the regression not too great, the fitted lines will be approximately parallel and not too misleading. The farther apart $\overline{X}_1$ and $\overline{X}_0$ are, the more different will be the slopes of the curve at the X values in the two groups, and the greater will be the difference in the estimated slopes of the two regression lines.

(The dashed lines in Figure 8.4 illustrate the two different linear regressions.) So, we will be faced with all the problems of nonparallel regression described above.

One way to handle suspected nonlinearity is to model the nonlinear regressions. By making a transformation of the X variable, we may obtain a much better fit to the observations. For example, we might find that a model of the form

$$Y = \alpha_i + \beta\sqrt{X} + e$$

adequately describes the data. A standard ANCOVA can then be carried out using $\sqrt{X}$ rather than X as the covariate.

The second case to consider is that of a nonlinear relationship between Y and X which is not necessarily parallel for the two groups, but which can be turned into a standard model by appropriate transformations of Y and/or X. For example, suppose that the model is given by

$$Y = \exp(\alpha_i + \beta X + e).$$

Then

$$\log Y = \alpha_i + \beta X + e$$

Thus using a logarithmic transformation on Y will allow the standard ANCOVA to be employed. Of course, it must be remembered after the analysis that the effect is defined in transformed (in this case, logarithmic) units. So it may be necessary to transform back to the original units in order to interpret the estimated effect. For example, suppose that an ANCOVA on the log scale produces an estimate

$$\alpha_1 - \alpha_0 = 3.$$

Then in terms of the original model, we have

$$Y = \begin{cases} \exp(\alpha_0 + \beta X + e) & \text{for treatment 0} \\ \exp(3 + \alpha_0 + \beta X + e) & \text{for treatment 1.} \end{cases}$$

Note that

$$\exp(3 + \alpha_0 + \beta X + e) = \exp(3)\exp(\alpha_0 + \beta X + e)$$

$$= (20.1)\exp(\alpha_0 + \beta X + e).$$

So the estimated effect of changing from treatment 0 to treatment 1 is to *multiply* the response by a factor of about 20.

Finding the appropriate transformations is largely a matter of trial and error. Standard statistical texts offer some guidance (see, e.g., Chatterjee and Price, 1977; Draper and Smith, 1966; Mosteller and Tukey, 1977; Tukey, 1977).

The third case involves nonlinear, nonparallel regressions where no suitable transformation can be found. In this case, both interpretational and technical problems become very difficult. Some recent research has been conducted on the comparison of quadratic regressions (Rogosa, 1977; Borich et al. 1976; Wunderlich and Borich, 1973), but in general the analyst can do no better than to fit separate regressions for the two groups.

8.5.3   Measurement Error

The third possible threat to the validity of ANCOVA is measurement error in the covariate. Classical measurement theory (see Lord and Novick, 1968) defines the *reliability* of a variable as the percentage of its variance attributable to variation in the true characteristic of interest. This notion is meaningful if we think of the observed score as the sum of true and error components. In Section 5.2 we discussed measurement error as a general issue in statistical adjustment.

Suppose that if we knew the true covariate scores, an ANCOVA model using them would accurately describe the data. Sometimes the equations relating outcomes to true (but unmeasurable) covariates are known as *structural equations*. If we use our imperfectly reliable, but observable, covariate, the resulting $\hat{\beta}$ turns out to be a biased estimate of the β in the structural equation. A biased treatment comparison will result, with the nature of the bias depending upon the nature of the measurement error. An appropriate correction is possible if the reliability of the covariate is known or can be estimated (see Cochran, 1968; DeGracie and Fuller, 1972; Lord, 1960; Stroud, 1974). However, these methods are quite complex and heavily dependent on certain untestable assumptions. So it is probably wiser to focus attention on collecting reliable information rather than trying to assess precisely the degree of reliability and adjust for it in the analyses.

8.5.4   Multiple Confounding Variables

Finally, we discuss the situation when other differences between groups in addition to those related to our measured covariate are suspected. If we can identify and measure other confounding variables, we can adjust for several covariates at once. Suppose, for example, that the model described by (8.5) holds. Then it is possible to obtain unbiased estimates $\hat{\beta}$ and $\hat{\gamma}$ of both β and γ to use these in our adjustment. For example, the adjustment treatment means would be given by

$$\overline{Y}_i - \hat{\beta}(\overline{X}_i - \overline{X}) - \hat{\gamma}(\overline{Z}_i - \overline{Z}).$$

Combining the ability to use transformations of the data with the capability for multivariate adjustments allows great flexibility in fitting an appropriate model for the data. This flexibility must, however, be weighed against the need to verify that all assumptions are met in this more complex situation.

We discussed above the problems in the single covariate situation resulting from possible differences in regression slopes, nonlinearity, and measurement error. With multiple covariates these problems are compounded. When several covariates are involved, we cannot use simple graphical methods to help in assessing the validity of assumptions, and models for measurement error become extremely complex.

The data analyst is faced with a dilemma. To obtain a good fit to the data for each group and include all potential confounding factors, he or she is tempted to include several covariates. But the more covariates included, the greater the potential problems in meeting and verifying the basic ANCOVA assumptions.

Now it might be though that the analyst should simply include the one or two most important possible confounding factors, expecting to eliminate most of the bias and avoiding the complexity of multiple covariates. While this procedure may often work well, there are situations where it can be quite misleading. It may even result in an estimate of treatment effect that is more biased than the unadjusted difference of group means. An artificial example of this phenomenon was given in Section 5.1. As another example of how this might occur, suppose that in the Greenberg (1953) data the rural children were not only older, but also tended to have shorter parents in such a way that the effects of these two factors, age and heredity, were exactly counterbalanced. Then, by using AN-COVA to adjust for age differences between groups, we would unwittingly create an artificial difference between the groups.

This example illustrates the care which is necessary in drawing inferences on the basis of ANCOVA. While a preponderance of short parents in one group might be an obvious factor to take into account, a confounding variable may be much more difficult to identify in other practical problems. It would be nice to give some simple guidelines for dealing with this problem. Unfortunately, there is no way to guarantee that the ANCOVA model is correctly specified. As with other statistical adjustment strategies considered in this book, the investigator may be criticized for omitting a particular confounding variable thought by someone else to be important. The general discussion of this problem contained in Chapter 5 includes some broad guidelines on choosing an adequate covariate set. A more detailed discussion of the issues in the context of ANCOVA is presented by Weisberg (1979), and some practical guidelines are offered by Cochran (1965).

APPENDIX 8A   FORMULAS FOR ANALYSIS-OF-COVARIANCE CALCULATIONS

We consider the general situation where K treatments are being compared. These will be indexed by $k = 1, 2, \ldots, K$. Let X_{ik} and Y_{ik} represent the covariate and outcome values for individual i in group k. Let $\overline{X}_k$ and $\overline{Y}_k$ be the means for the n_k individuals in group k. Then we can define the between-group (treatment) sums of squares and cross-products by

$$T_{xx} = \sum_{k=1}^{K} n_k (\overline{X}_k - \overline{X})^2$$

$$T_{yy} = \sum_{k=1}^{K} n_k (\overline{Y}_k - \overline{Y})^2$$

$$T_{xy} = \sum_{k=1}^{K} n_k (\overline{X}_k - \overline{X})(\overline{Y}_k - \overline{Y}),$$

where $\overline{X}$ and $\overline{Y}$ are the grand means of X and Y across all groups. Similarly, we define within-group (error) sums of squares and cross-products:

$$E_{xx} = \sum_{k=1}^{K} \sum_{i} (X_{ik} - \overline{X}_k)^2$$

$$E_{yy} = \sum_{k=1}^{K} \sum_{i} (Y_{ik} - \overline{Y}_k)^2$$

$$E_{xy} = \sum_{k=1}^{K} \sum_{i} (X_{ik} - \overline{X}_k)(Y_{ik} - \overline{Y}_k),$$

where Σ_i indicates the sum over individuals within each group. We also define the quantity

f = total number of subjects minus number of groups

and, using the definitions above we define

$$S_{xx} = T_{xx} + E_{xx}$$

$$S_{yy} = T_{yy} + E_{yy}$$

$$S_{xy} = T_{xy} + E_{xy}.$$

Then we can calculate the residual mean squares for treatments and error:

$$s_e^2 = \left(E_{yy} - \frac{E_{xy}^2}{E_{xx}}\right) \bigg/ (f - 1)$$

$$s_t^2 = \left(T_{yy} - \frac{S_{xy}^2}{S_{xx}} + \frac{E_{xy}^2}{E_{xx}}\right) \bigg/ (K - 1).$$

These can be used to calculate an F statistic to test the null hypothesis that all treatment effects are equal:

$$F = \frac{s_t^2}{s_e^2}.$$

Under the null hypothesis this ratio has an F distribution with $K - 1$ and $f - 1$ degrees of freedom. The estimated regression coefficient of Y on X is

$$\hat{\beta} = \frac{E_{xy}}{E_{xx}}.$$

From the definitions of E_{xy} and E_{xx} given above, it is clear why this is called a pooled within-group estimator. The estimated standard error for the adjusted difference between two group means (say group 0 and group 1) is given by

$$s_d = s_e \sqrt{\frac{1}{n_0} + \frac{1}{n_1} + \frac{(\overline{X}_1 - \overline{X}_0)^2}{E_{xx}}},$$

when n_0 and n_1 are the sample sizes of the two groups. A test of the null hypothesis that the adjusted difference is zero is provided by the statistic

$$t = \frac{\overline{Y}_1 - \overline{Y}_0 - \hat{\beta}(\overline{X}_1 - \overline{X}_0)}{s_d}.$$

Under the null hypothesis, it has a t distribution with $f - 1$ degrees of freedom.

REFERENCES

Belson, W. A. (1956), A Technique for Studying the Effects of a Television Broadcast, *Applied Statistics*, **5**, 195–202.

Borich, G.D., Godbout, R. C., and Wunderlich, K. W. (1976), *The Analysis of Aptitude-Treatment Interactions: Computer Programs and Calculations*, Austin, TX: Oasis Press.

Chatterjee, S., and Price, B. (1977), *Regression Analysis by Example*, New York: Wiley.

Cochran, W. G. (1957), Analysis of Covariance: Its Nature and Uses, *Biometrics*, **13**, 261–281.

Cochran, W. G. (1965). The Planning of Observational Studies of Human Populations, *Journal of the Royal Statistical Society, Series A*, **128**, 234–266.

Cochran, W. G. (1968), Errors of Measurement in Statistics, *Technometrics*, **10**(4), 637–666.

Cochran, W. G. (1969), The Use of Covariance in Observational Studies, *Applied Statistics*, **18**, 270–275.

Cochran, W. G., and Rubin, D. B. (1973). Controlling Bias in Observational Studies: A Review, *Sankhya, Series A*, **35**(4), 417–446.

Cronbach, L. J., Rogosa, D. R., Floden, R. E., and Price, G. G. (1977) Analysis of Covariance in Nonrandomized Experiments: Parameters Affecting Bias, occasional paper of the Stanford Evaluation Consortium.

DeGracie, J. S., and Fuller, W. A. (1972), Estimation of the Slope and Analysis of Covariance When the Concomitant Variable Is Measured with Error, *Journal of the American Statistical Association*, **67**, 930–937.

Draper, N. R., and Smith, H. (1966), *Applied Regression Analysis*, New York: Wiley.

Elashoff, J. D. (1969), Analysis of Covariance, a Delicate Instrument, *American Educational Research Journal*, **6**, 383–402.

Fisher, R. A. (1932), *Statistical Methods for Research Workers,* 4th ed., Edinburgh: Oliver & Boyd.

Glass, G. V., Peckham, P. D., and Sanders, J. R. (1972), Consequences of Failure to Meet Assumptions Underlying the Analysis of Variance and Covariance, *Review of Educational Research,* **42,** 237–288.

Greenberg, B. G. (1953), The Use of Analysis of Covariance and Balancing in Analytical Surveys, *American Journal of Public Health,* **43**(6), 692–699.

Johnson, P. O., and Neyman, J. (1936), Tests of Certain Linear Hypotheses and Their Applications to Some Educational Problems, *Statistical Research Memoirs,* **1,** 57–93.

Lord, F. M. (1960), Large-Scale Covariance Analysis When the Control Variable Is Fallible, *Journal of the American Statistical Association,* **55,** 307–321.

Lord, F. M., and Novick, M. R. (1968), *Statistical Theories of Mental Test Scores,* Reading, MA: Addison-Wesley.

Mosteller, F. C., and Tukey, J. W. (1977), *Data Analysis and Regression,* Reading, MA: Addison-Wesley.

Potthoff, R. F. (1964), On the Johnson–Neyman Technique and Some Extensions Thereof, *Psychometrika,* **29,** 241–256.

Rogosa, D. R. (1977), Some Results for the Johnson–Neyman Technique, unpublished doctoral dissertation, Stanford University.

Stroud, T. W. F. (1974), Comparing Regressions When Measurement Error Variances Are Known, *Psychometrika,* **39**(1), 53–68.

Tukey, J. W. (1977), *Exploratory Data Analysis,* Reading, MA: Addison-Wesley.

Walker, H. M., and Lev, J. (1959), *Elementary Statistical Methods,* New York: Holt, Rinehart and Winston.

Weisberg, H. I. (1979), Statistical Adjustments and Uncontrolled Studies, *Psychological Bulletin,* **86,** 1149–1164.

Winer, B. J. (1971), *Statistical Principles in Experimental Design,* 2nd ed., New York: McGraw-Hill.

Wunderlich, K. W., and Borich, G. D. (1973), Determining Interactions and Regions of Significance for Curvilinear Regressions, *Educational and Psychological Measurement,* **33,** 691–695.

CHAPTER 9

Logit Analysis

Logit analysis can be applied in comparative studies to estimate the effect of a risk factor on a dichotomous outcome factor as measured by the odds ratio. The usefulness of logit analysis is its ability to adjust for many confounding variables simultaneously. These confounding variables can be either categorical or numerical.

We will begin by motivating the mathematical model that underlies logit analysis (Section 9.1) and showing how logit analysis can be used to control for a confounding variable (Section 9.2). Details of various aspects of implementation are given in Sections 9.3 to 9.8 with some additional mathematical details in Appendix 9A. Initially, discussion is restricted to cohort studies; the differences applicable to case-control studies are presented in Section 9.6.

9.1 DEVELOPING THE LOGIT ANALYSIS MODEL

Consider the problem of determining whether diabetes is a contributing (risk) factor to heart disease. To keep things simple, suppose that heart disease, the outcome factor, is a dichotomous variable (present or absent) as is the risk factor, diabetes (also present or absent). The confounding factors to be controlled for in this problem include the categorical variable sex and the numerical variables age, diastolic blood pressure, and serum cholesterol level.

One technique that could be applied to this problem is stratification (Chapter 7), but stratification requires that the confounding variables as well as the risk variable be categorical. This means that the numerical confounding variables would have to be categorized. (For example, blood pressure could be changed to a two category variable with categories high and not high.) As discussed in Chapter 7, this process generally leads to many strata and a resulting lowering of precision of the estimated effect. Moreover, it can only approximately control for the confounding variables; for example, persons whose blood pressure is just low enough so as to just miss being classified as having high blood pressure are treated the same as persons with much lower blood pressures. For these reasons it is desirable to have an adjustment technique that can deal with confounding factors such as blood pressure in their numerical form. Logit analysis does this.

For ease of exposition, we will begin with a single confounding variable, and, for the time being, we exclude the case of a categorical confounding variable with more than two categories. The generalizations to multiple confounding variables and to categorical confounding variables with more than two categories are discussed in Section 9.8.

Logit analysis is analogous to linear regression analysis or, more specifically, to analysis of covaraince (Chapter 8); in many cases logit analysis will accomplish for comparative studies with categorical outcome variables what analysis of covariance can accomplish for comparative studies with numerical outcomes. Specifically, both yield estimates of the treatment effect adjusted for a confounding variable or variables. As we will see, issues such as parallelism that are important for analysis of covariance are also important for logit analysis.

Linear regression is based on the assumption that the mean of the distribution of the response or outcome variable, Y, is a linear combination of the background (or "independent") variables. For our applications the independent variables are the risk and confounding variables. For a single confounding variable X, the linear regression model is the analysis-of-covariance model (8.1):

$$Y = \alpha_0 + \beta X + \epsilon \qquad (9.1a)$$

if the risk factor is absent and

$$Y = \alpha_1 + \beta X + \epsilon \qquad (9.1b)$$

if the risk factor is present. Equations 9.1a and 9.1b can be rewritten as a single equation by introducing a variable, R, for the risk factor, where $R = 1$ indicates the risk factor is present and $R = 0$ indicates the risk factor is absent. The regression equations (9.1a and 9.1b) then become

Linear regression model: $Y = \alpha_0 + \tau R + \beta X + \epsilon,$ $\qquad (9.2)$

where $\tau = \alpha_1 - \alpha_0$.

If Y were a numerical variable, the analysis of covariance, based on (9.2), would be the proper means of estimating τ as a measure of the effect of the risk factor after controlling for the confounding variable X. If, instead, the outcome is dichotomous, such as heart disease present–heart disease absent, Y would be a dichotomous variable taking the values 1 (heart disease present) and 0 (heart disease absent). With such a dependent variable, the linear regression approach would not be generally appropriate, for two reasons. First, the regression model (9.2) says that the mean of the distribution of Y for given values of X and R, denoted by $E[Y|R, X]$, is $\alpha_0 + \tau R + \beta X$. For a dichotomous Y, the mean of Y is bounded by 0 and 1, but $\alpha_0 + \tau R + \beta X$ is not. Therefore, the regression model can be at best an approximation over a limited range of X values, since it is possible to obtain estimates of $\alpha_0 + \tau R + \beta X$ that are less than 0 or greater than 1. Second, the assumption of normally distributed error terms that is necessary for hypothesis testing is not even plausible, since the error term, ϵ, is itself dichotomous, taking on the values

$$-E[Y|R, X] \quad \text{if } Y = 0 \qquad \text{and} \qquad 1 - E[Y|R, X] \quad \text{if } Y = 1.$$

An alternative to linear regression is to say that some function of the mean of Y is a linear combination of X and R. If we let $P(R, X)$, or just P, denote the probability that $Y = 1$ given the values of the risk variable R and the confounding variable X, then $E[Y|R, X] = P(R, X)$. We will now consider functions of $P(R, X)$. A common choice for this function is the log odds, also known as the logit transform of P. The choice of the logit transform gives us what is known as the *logit model*:

$$\ln\left[\frac{P(R, X)}{1 - P(R, X)}\right] = \alpha + \gamma R + \beta X. \qquad (9.3)$$

[The reason for replacing the τ of (9.2) with γ in (9.3) will be made clear in Section 9.2.] The quantity on the left-hand side of (9.3) is the logit transform of P, commonly denoted by logit(P). The logit transform has the desirable property of transforming the (0, 1) interval for P to $(-\infty, +\infty)$, so that there is no longer any concern about $\alpha + \gamma R + \beta X$ lying outside the unit interval.

The analysis-of-covariance model (9.1) for a numerical X corresponds to two parallel lines, $\alpha_0 + \beta X$ and $\alpha_1 + \beta X$, as in Figure 8.1. The logit model (9.3) also corresponds to two parallel lines, $\alpha + \beta X$ and $\alpha + \gamma + \beta X$, but in a transformed scale (Figure 9.1).

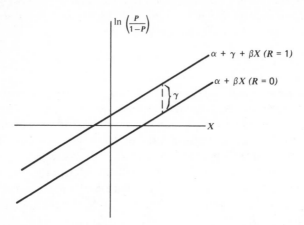

Figure 9.1 Logit model in transformed (logit) scale.

In the untransformed, P, scale, the two straight lines from Figure 9.1 would appear as shown in Figure 9.2. The S-shaped curves in Figure 9.2 indicate the effect of the logit transform. In their center, around $P = 0.5$, the curves are nearly straight lines and the logit transform has little effect there. At either extreme the curves flatten out. The implication is that the change in X required to change P from 0.98 to 0.99, say, is much greater than the change in X required to change P from 0.49 to 0.50.

There are, in general, no theoretical justifications for the form (9.3) (with the exception of some particular distributional assumptions that are discussed in Section 9.4). Many other choices for a transformation of P are possible, the most common being the inverse normal cumulative distribution function, leading to probit analysis; see, for example, Finney (1964) and Goldberger (1964). As discussed by Hanushek and Jackson (1977, Chap. 7), there is little practical difference between the logit and probit curves. The principal appeal of the logit is the ease, relative to the probit and other choices of transformations, with which the estimates of the parameters may be found. The fact that logit analysis has been found to be useful in many fields of application for both comparative studies (Cornfield, 1971, for example) and other uses, such as predicting a commuter's choice of travel mode (Stopher, 1969; Talvitie, 1972), is no reason to expect it to be a universally good approximation to the relationship between P and the variables X and R. (Further discussion of this issue can be found in Gordon, 1974.) Some methods for checking the appropriateness of a particular logit model are discussed in Section 9.7.

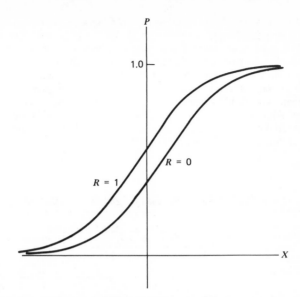

Figure 9.2 Logit model in untransformed (P) scale.

9.2 USE OF LOGIT ANALYSIS TO CONTROL FOR A CONFOUNDING VARIABLE

The problem is to estimate the effect associated with a risk factor, controlling for a confounding factor. When using logit analysis, the odds ratio (Chapter 3) is usually chosen as the measure of the treatment effect, since, as we will see, it is very simply related to the parameters of the logit model (9.3). The logit model for the probability, P, that the outcome is a "success" is given by (9.3). (Following common practice, the two possible outcomes will be referred to as "success" and "failure.") The odds ratio is then

$$\psi = \frac{P(1, X)}{1 - P(1, X)} \bigg/ \frac{P(0, X)}{1 - P(0, X)} = \exp(\gamma)$$

independent of X. We may thus identify γ as the (natural) logarithm of the odds ratio (so this use of γ is consistent with other chapters). Logit analysis yields an estimate $\hat{\gamma}$ of γ and then estimates the odds ratio, ψ, by $\hat{\psi} = \exp(\hat{\gamma})$.

The logit model (9.3) implies that the odds ratio is constant over all values of the confounding variable X. This is the usual no-interaction, or parallelism, assumption (Section 3.3), corresponding to the two parallel lines in Figure 9.1. In Section 9.7 we will consider how to check the validity of the no-interaction assumption.

Example 9.1   Analysis of data from Head Start Program: As an example, data from the Head Start Program (Smith, 1973; Weisberg, 1973) will be used. The data are from a cohort study consisting of two groups of preschool children—one group consisting of children who had taken part in Head Start, the other of children who had not. The risk variable R is defined to be 1 for children in the Head Start group and 0 for the other children. The outcome of interest is whether or not the child has attained minimal competency in general knowledge at the time of entry to school as measured by the Preschool Inventory Test. The outcome variable Y is defined to be 1 if the child had attained minimal compentency and 0 if not. The results, ignoring any possible confounding, for a total of 468 children in this study are shown in Table 9.1.

Table 9.1   Head Start Data, Ignoring Confounding

Outcome Factor	Risk Factor		Total
	Not Head Start	Head Start	
Not minimally competent	274	61	335
Minimally competent	97	36	133
Total	371	97	468

The estimate of the odds ratio without controlling for any confounding, denoted by $\hat{\psi}^C$, where the superscript C stands for crude, is given by

$$\hat{\psi}^C = \frac{(274)(36)}{(97)(61)} = 1.667,$$

indicating that the odds of attaining minimal competancy are approximately 1.7 times higher for children who had participated in the Head Start Program than for those who had not.

One important confounding variable in this study is the years of education of the head of household, a numerical variable. Letting X denote years of education, an estimate of the odds ratio controlling for years of education is found by fitting the logit model (9.3). This yields

$$\hat{\alpha} = -2.931$$
$$\hat{\gamma} = 0.325$$
$$\hat{\beta} = 0.186.$$

The logit analysis estimate of ψ is then found as

$$\hat{\psi} = \exp(0.325) = 1.384.$$

The estimate of β is a measure of the effect of the confounding variable on the outcome. In this case it indicates that each additional year of education of the head of household is estimated to increase the log odds of a minimally competent child by 0.186 [or multiplied the odds of a success by $\exp(0.186) = 1.204$].

The α in the logit model is like the intercept in the equation for a straight line. The value of α will depend on how the variables are defined. Consequently, the value of α or $\hat{\alpha}$ is not generally interpretable. The quantity $\hat{\alpha}$ is required for estimating risks (probabilities), which is done by substituting the estimates of α, γ, and β into (9.3) and solving for $P(R, X)$.

9.3 PARAMETER ESTIMATION BY MAXIMUM LIKELIHOOD

The standard procedure for estimating the parameters of the logit model (9.3) is maximum likelihood. This requires an iterative procedure. The mathematical and some computational details are discussed in Appendix 9A. The key assumption is that the responses or outcomes are observations of independent binomial (or Bernoulli) variables. If there is dependence between the observations such as with matched samples, the usual logit analysis is not appropriate. (Logit analysis of matched samples is discussed in Section 13.3.2.)

The known properties of the maximum likelihood estimates are based on large-sample theory. In large samples the maximum likelihood estimates will be approximately unbiased. The bias of parameter estimates from logit analysis in small samples is not known.

When all the confounding variables as well as the risk and outcome variables are categorical, then, as shown by Bishop (1969), logit analysis is a special case of log-linear analysis (Chapter 10). Maximum likelihood estimation of the appropriate log-linear model would give the same estimates as maximum likelihood estimation of the logit model (9.3). (Presentation of the relation between the two models is postponed to Section 10.6.) However, the numerical procedures for finding the estimates differ for the two methods. So when all the factors are categorical, one can choose logit or log-linear analysis, depending on which is easier to use for the problem of interest. The choice will depend on the computer programs available as well as the statistics desired. Parameter estimates tend to be more easily obtained from logit analysis, whereas hypothesis testing can be much easier with log-linear analysis, particularly when the number of confounding variables is large and there are many interactions between the risk variable and the confounding variables.

9.4 OTHER PARAMETER ESTIMATION PROCEDURES

There are two other estimation procedures that can be used for particular purposes and applications as alternatives to maximum likelihood. The most important of these is linear discriminant analysis. (See Snedecor and Cochran,

1967, Chap. 13, for a discussion of discriminant analysis.) Linear discriminant analysis assumes that there are two populations, denoted by $Y = 1$ and $Y = 0$, and that, within each population, X and R have a bivariate normal distribution, with the variances and covariances identical in the two populations, but different means. It can be shown (Welch, 1939) that these distributional assumptions lead to the form (9.3) for the conditional distribution of Y given X and R, where the logit parameters, α, γ, and β, are functions of the normal distribution parameters. [The form (9.3) can also be obtained for distributions other than the normal.] The normal distribution parameters are easy to estimate in the usual manner using the sample means, variances, and covariances. Estimates of the logit parameters are then found by substitution into the appropriate functions. See Cornfield et al. (1961), Cornfield (1962), and Truett et al. (1967) for the details of this procedure.

Estimating the logit parameters based on the discriminant model is much easier than the maximum likelihood approach. For this reason many people prefer it to maximum likelihood, even when the assumption of normally distributed X and R is known to be violated, as would be the case, for example, when the risk factor is dichotomous. Halperin et al. (1971) and Seigel and Greenhouse (1973) discuss the consequences of using the linear discriminant approach when the normality assumption is violated. Essentially, the results are that the discriminant estimates of the odds ratio will remain biased in large samples. In terms of bias, maximum likelihood based on (9.3) is then preferable to the linear discriminant function approach unless the normality assumption is satisfied. On the other hand, when the normality assumption is appropriate, Efron (1975) has shown that the discriminant estimates will be more precise than estimates obtained by maximum likelihood based directly on (9.3). Presumably, the advantages of the discriminant function estimates will hold for small departures from the normality assumption, but it is not known for how much of a departure this will be true.

When the normality assumption is untenable, the discriminant function approach can still have some value, even though the parameter estimates will be biased. First, Gilbert (1968) and Moore (1973) have shown that the discriminant approach is still reasonable for discrimination purposes, that is, for predicting the outcome for given values of the risk and confounding variables. [This may just reflect that Fisher's (1936) derivation of the linear discriminant function as a discriminator and not as a means of estimating logit parameters did not require normality.] Halperin et al. (1971) have also shown that the discriminant function works reasonably well for isolating important variables and so could be used to determine which of a set of confounding variables are important enough to include in the analysis. Some researchers have also found that, for their particular data, the parameter estimates from discriminant analysis were similar to those from logit analysis—Kahn et al. (1971) and Talvitie (1972). The implication of all this is that discriminant analysis, which is computationally easier than maximum likelihood logit analysis, can be used as a variable selection procedure before using maximum likelihood.

This variable selection procedure can be simplified. As shown by Fisher (1936) (also see Anderson, 1958, Sec. 6.5.4), the ordinary (unweighted) linear regression of the dichotomous Y on X and R yields coefficients that are proportional to the linear discriminant function coefficients. (Regression with a 0–1 dependent variable, as here, is frequently referred to as binary regression.) This means that simple linear regression can be used in place of discriminant analysis as a means for selecting important confounding variables to be included in a logit analysis.

The second alternative estimation procedure is applicable only if the subjects in the study fall into moderately large groups when classified by X and R (i.e., all members of a group have the same values of X and R). This will generally require a categorical X but could also apply if there were replications at each observed value of a numerical X. Then there is a noniterative linear regression procedure, frequently referred to as logistic regression, which will give approximately the same results as maximum likelihood in large samples. An outline of this procedure follows; see Hanushek and Jackson (1977, Sec. 7.4) or Theil (1970) for a more complete discussion.

Let $N(R, X)$ be the total number of subjects in the group determined by X and R, and let $p(R, X)$ be the proportion of those subjects whose outcome is a "success." The procedure is to fit the regression model

$$\ln\left[\frac{p(R, X)}{1 - p(R, X)}\right] = \alpha + \gamma R + \beta X + \epsilon. \tag{9.4}$$

[The quantity to the left of the equals sign in (9.4) is the natural logarithm of the sample odds, sometimes referred to as the empirical logistic transform.] Since the error terms in this model do not have equal variances, weighted regression is necessary with weights

$$W = N(R, X)p(R, X)[1 - p(R, X)].$$

[See Mosteller and Tukey (1977, Chap. 14), for example, for a discussion of weighted regression.] If X is a categorical variable with more than two categories, the method of Section 9.8 must be applied to (9.4).

A modification of the left side of (9.4) is necessary to handle cases where the proportion, $p(R, X)$, is 0 or 1. Berkson (1955) presented one such modification and a corresponding modification to the weight W. Based on the work regarding estimation of the log odds ratio (Section 7.6 and Appendix 7B), a better modification would be (7.12). Cox (1970, Chap. 3) discusses the later modification to the empirical logistic transform and the corresponding choice of weights. The effect of the choice of modification to (9.4) becomes negligible in large samples.

In bioassay, the logistic regression procedure is called the minimum logit chi-square estimator and was originally proposed by Berkson (1944) (see Ashton, 1972). No study of the minimum logit chi-square estimator has been done for more than one independent variable and hence that would apply to the case of interest here (9.4). Still, the available results indicate that this procedure is a reasonable choice where appropriate. (Berkson (1955) and Hauck (1975) have shown, for the case of a single numerical independent variable [(9.4) with $\gamma = 0$], that the minimum logit chi-square estimator of the slope β is generally more precise and less biased than the maximum likelihood estimator in small samples.) The appropriateness issue is important. Forming groups by stratifying a numerical confounding variable, for example, would not be appropriate, since that would be equivalent to errors of measurement in the independent variables in regression analysis, with the consequence that the minimum logit chi-square estimator would remain biased in large samples instead of becoming equivalent to the maximum likelihood estimator.

9.5   HYPOTHESIS TESTING

To test whether the treatment effect, here measured by the log of the odds ratio, is different from zero, two different large-sample tests are commonly used. The first is a likelihood ratio test. [See Hoel (1962, Chap. 9) or similar texts for a discussion of likelihood ratio tests.] This requires fitting the logit model (9.3) twice, once to estimate all three parameters and a second time to estimate α and β when γ is restricted to be 0.

The second test is due to Wald (1943). This test, "Wald's test," has the advantage of being easier to compute. It is calculated in a manner similar to common t statistics, namely by dividing the estimate of γ by the standard error of that estimate. (See Appendix 9A for the calculation of the standard error.) The distribution of the resulting statistic will be approximately standard normal in large samples when the null hypothesis, $\gamma = 0$, is true. Even though Wald's test and the likelihood ratio test become equivalent in sufficiently large samples and Wald's test is easier to compute, the likelihood ratio test is preferable. Hauck and Donner (1977) found that, when applied to logit analysis, Wald's test can behave in an aberrant manner, yielding apparently insignificant results when the likelihood ratio test strongly suggests rejecting the null hypothesis.

9.6   CASE-CONTROL STUDIES

Logit analysis can be applied to case-control studies in either of two ways. The first is to interchange the role of the risk and outcome factors in (9.3), that is, to work with the probability of being in the risk factor present group given the observed outcome. (This is assuming a dichotomous risk factor.) Let $P'(Y, X)$ denote the probability that the risk factor is present given the values of the outcome, Y, and the confounding variable, X. Then the logit model would be

$$\ln\left[\frac{P'(Y, X)}{1 - P'(Y, X)}\right] = \alpha' + \gamma Y + \beta'X. \qquad (9.5)$$

The log of the odds ratio, γ, can then be estimated by maximum likelihood just as for cohort studies.

The only difficulty with using (9.5) is that it is not natural to think of probabilities of belonging to risk factor categories given the outcome. The model (9.3), on the other hand, is based on the more natural approach of treating the outcome as dependent on the risk factor. Mantel (1973) has shown that it is permissible to use the cohort model (9.3) even though the study is case-control, as long as the sample cases and controls are representative of the population cases and controls in the sense that the probability of being selected from the population cases (or controls, respectively) is independent of the risk and confounding factors. (Otherwise, there is a selection bias, as discussed in Section 4.5.2.) The estimates of γ and β will be unaffected. The only negative consequence is that the estimate of α will be biased by an amount that depends on the sampling fractions for the cases and controls. This later fact simply reflects, in terms of this particular statistical procedure, the general fact that risks $[P(R, X)]$ are not estimable from case-control studies. However, if the two sampling fractions are known, the estimate of α can be corrected, making it possible to estimate risks.

Breslow and Powers (1978) show that only if X (and R) are categorical variables and the categories of X are represented in the model using the method of Section 9.8 do the two approaches, (9.3) and (9.5), yield the same estimate of the odds ratio. Otherwise, the two approaches yield different estimates.

9.7 CHECKING THE MODEL

There are two questions to be answered in checking the model. The first is the general question of whether the model (9.3) fits the data well enough to serve as a basis for estimating the treatment effect. The second is the specific problem of checking whether the parallelism, or no-interaction, assumption holds, that is, checking whether the odds ratio is constant over all values of the confounding variable. To answer the first question requires a goodness-of-fit test. In those applications of logit analysis where log-linear analysis is also appropriate (i.e. where the data are in the form of a contingency table), usual goodness-of-fit statistics, Pearson and likelihood chi-squares, can be applied. See Section 10.4.2 for goodness-of-fit testing in such cases.

If it is suspected or found that the model (9.3) does not fit well, the fit can sometimes be improved in a manner similar to that used in linear regression analysis, namely by the addition of other confounding variables. The added variables can be either new variables or transformations, such as polynomial terms, of the old ones. For example, blood pressure could be replaced by log (blood pressure).

The question of checking for interaction is easier to handle. If the interaction between the risk and the confounding variables is such that the logarithm of the odds ratio is linearly related to the confounding variable (i.e., has the form $\gamma + \lambda X$), then the corresponding logit model is

$$\ln\left[\frac{P(R, X)}{1 - P(R, X)}\right] = \alpha + \gamma R + \beta X + \lambda RX. \tag{9.6}$$

To test for interaction of this form it is only necessary to fit the four-parameter model (9.6) in addition to the usual model (9.3) and test the null hypothesis $\lambda = 0$ using a likelihood ratio test.

Example 9.2   Testing for interaction in Head Start Program data: Consider Example 9.1 and the problem of determining whether the odds ratio varies with level of head of household's education. Fitting model (9.6) yields

$$\hat{\alpha} = -2.609$$
$$\hat{\gamma} = -2.522$$
$$\hat{\beta} = 0.155$$
$$\hat{\lambda} = 0.253.$$

The estimated logarithm of the odds ratio is now $-2.522 + 0.253$ (years of education of head of household). The likelihood ratio test for testing the null hypothesis of no interaction ($\lambda = 0$) yields a one degree-of-freedom chi-square of 2.542 ($P = 0.111$), so the null hypothesis is not rejected at the 11% level.

9.8   MULTIPLE CONFOUNDING FACTORS

For ease of exposition, the discussion to this point has assumed a single confounding variable. The great usefulness of logit analysis, however, is that it can handle any number of confounding variables.

Suppose that there are K confounding variables $X_1, \ldots, X_K$. (These K may depend on some smaller number of factors by including transformations, product terms, and the like.) The logit model then becomes

$$\ln\left[\frac{P(R, X_1, \ldots, X_K)}{1 - P(R, X_1, \ldots, X_K)}\right] = \alpha + \gamma R + \sum_{k=1}^{K} \beta_k X_k. \tag{9.7}$$

Everything said in this chapter about the single-confounding-variable case applies as well to the multiple variable case. In particular, γ will still be the log of the odds ratio and estimation and hypothesis testing are done in the same way.

Example 9.3 *Analysis of Head Start Program data with two confounding variables:* Consider again the Head Start data of Example 9.1 and let us now suppose that we wish to control in addition for confounding due to race. Letting

$$X_1 = \text{years of education of head of household}$$

$$X_2 = \text{race } (1 = \text{black and } 0 = \text{white}),$$

the right side of (9.7) becomes

$$\alpha + \gamma R + \beta_1 X_1 + \beta_2 X_2.$$

The estimated parameters are

$$\hat{\alpha} = -2.388$$

$$\hat{\gamma} = -0.019$$

$$\hat{\beta}_1 = 0.178$$

$$\hat{\beta}_2 = -0.850.$$

Adjusting for both confounding variables, the estimate of the odds ratio is $\exp(-0.019) = 0.981$.

One situation that leads to multiple confounding variables even though there is only one confounding factor is if the one factor is categorical with more than two categories. This confounding factor must be represented in the logit model by a number of indicator (or dummy) variables. If there are m categories to the confounding factor, $m - 1$ dichotomous (0–1) indicator variables, $X_1, \ldots, X_{m-1}$ are necessary. Each X_k corresponds to one category of the confounding factor. They are defined by

$$X_k = \begin{cases} 1 & \text{if the subject is in category } k \text{ of the confounding factor} \\ 0 & \text{otherwise.} \end{cases}$$

The β_k corresponding to X_k is then the increase in the log odds of a "successful" outcome given that the subject is in category k rather than category m of the confounding factor. (Category m is the category without a corresponding indicator variable.)

The numbering of the categories of the confounding factor is arbitrary, so, statistically, any category can be specified as the mth. The mth category ought to be chosen, however, to make the coefficients (the β_k's) as interpretable and useful as possible. For example, suppose that the confounding factor is employment status with three categories—unemployed, employed part-time, employed full-time. A reasonable choice for indicator variables would be:

$$X_1 = \begin{cases} 1 & \text{if employed part-time} \\ 0 & \text{if unemployed or employed full-time} \end{cases}$$

$$X_2 = \begin{cases} 1 & \text{if employed full-time} \\ 0 & \text{if unemployed or employed part-time.} \end{cases}$$

β_1 (corresponding to X_1) would then measure the increase in the log odds due to being employed part-time rather than unemployed; β_2 would measure the increase due to being employed full-time rather than unemployed.

In the case of a single confounding factor that is categorical, the methods of stratification given in Section 7.6 are also applicable. The maximum likelihood estimate of the logarithm of the odds ratio mentioned there is the estimate that would be obtained by logit analysis of the model that included the appropriate indicator variables for the confounding factor categories.

APPENDIX 9A DETAILS OF THE MAXIMUM LIKELIHOOD APPROACH TO LOGIT ANALYSIS

In this appendix we will present some of the mathematical details of the maximum likelihood approach to logit analysis. This begins with the likelihood and likelihood equations and includes some of the details for calculating standard errors. Again, for ease of presentation, results are restricted to a single confounding variable.

As mentioned earlier, obtaining the maximum likelihood estimates requires an iterative procedure. The choice and method of implementation of an appropriate algorithm is a numerical analysis problem and will not be considered here. References on these topics include Walker and Duncan (1967), and Jones (1975).

9A.1 Likelihood Equations and Standard Errors

Let Y, R, and X be defined as elsewhere in this chapter and let the subscripts i and j donate the ith subject in the jth risk factor group ($i = 1, \ldots, n_j$ and $j = 0, 1$). We then have that $P_{ij} = P(R_{ij}, X_{ij})$, the probability of a "sucess" for the ith subject in the jth group, satisfies

$$\ln\left(\frac{P_{ij}}{1 - P_{ij}}\right) = \alpha + \gamma R_{ij} + \beta X_{ij}. \tag{9.8}$$

The basic assumption is that all the responses are independent. Consequently, the likelihood, L, is

$$L = \prod_{j=0}^{1} \prod_{i=1}^{n_j} P_{ij}^{Y_{ij}} (1 - P_{ij})^{1-Y_{ij}}.$$

Taking natural logarithms and substituting (9.8) the log likelihood, l, is

$$l = \sum_{j=0}^{1} \sum_{i=1}^{n_j} \{ Y_{ij}(\alpha + \gamma R_{ij} + \beta X_{ij}) - \ln[1 + \exp(\alpha + \gamma R_{ij} + \beta X_{ij})] \}. \quad (9.9)$$

Taking partial derivatives of l with respect to the three parameters α, γ, and β, we obtain the likelihood equations

$$\sum_{j=0}^{1} \sum_{i=1}^{n_j} Y_{ij} = \sum_{j=0}^{1} \sum_{i=1}^{n_j} \hat{P}_{ij}$$

$$\sum_{j=0}^{1} \sum_{i=1}^{n_j} R_{ij} Y_{ij} = \sum_{j=0}^{1} \sum_{i=1}^{n_j} R_{ij} \hat{P}_{ij} \quad (9.10)$$

$$\sum_{j=0}^{1} \sum_{i=1}^{n_j} X_{ij} Y_{ij} = \sum_{j=0}^{1} \sum_{i=1}^{n_j} X_{ij} \hat{P}_{ij}.$$

The maximum likelihood estimates, $\hat{\alpha}$, $\hat{\beta}$, and $\hat{\gamma}$ are the values of α, β, and γ that satisfy (9.10), and $\hat{P}_{ij}$ is P_{ij} with the three parameters replaced by their maximum likelihood estimates.

A problem that can occur is the existence of solutions to (9.10) where one or more parameter estimates are infinite. This occurs when the variables included in the logit model are perfect predictors (or discriminators) of the outcome. Actually, it is a generalization of the fact that the maximum likelihood estimate of the logarithm of the odds ratio from a 2×2 table is infinite if any one of the four table entries is zero. This is not a common problem except when sample sizes are small or one of the two outcomes is very rare. Though the point estimates and their standard errors will not be meaningful in such cases, likelihood ratio tests may still be done.

Standard errors are based on large-sample maximum likelihood theory that relates the variance of the estimates to the expected values of the second derivatives of the log likelihood. In practice, these variances can be estimated from the second derivatives of the sample log likelihood (9.9). The formulas for the standard errors depend on all six second derivatives, in the form of an inverse of a 3×3 matrix, so will not be given here.

REFERENCES

Anderson, T. W. (1958), *An Introduction to Multivariate Statistical Analysis*, New York: Wiley.

Ashton, W. D. (1972), *The Logit Transformation, with Special Reference to Its Use in Bioassay*, London: Charles Griffin.

Berkson, J. (1944), Application of the Logit Function to Bioassay, *Journal of the American Statistical Association*, **39**, 357–365.

Berkson, J. (1955), Maximum Likelihood and Minimum χ^2 Estimates of the Logistic Function, *Journal of the American Statistical Association*, **50**, 130–162.

Bishop, Y. M. M. (1969), Full Contingency Tables, Logits and Split Contingency Tables, *Biometrics*, **25**, 383–399.

Breslow, N., and Powers, W. (1978), Are There Two Logistic Regressions for Retrospective Studies? *Biometrics*, **34**, 100–105.

Cornfield, J. (1962), Joint Dependence of Risk of Coronary Heart Disease on Serum Cholesterol and Systolic Blood Presssure: A Discriminant Function Analysis, *Federation Proceedings*, **21** (suppl. 11), 58–61.

Cornfield, J. (1971), The University Group Diabetes Program: A Further Statistical Analysis of the Mortality Findings, *Journal of the American Medical Association*, **217**, 1676–1687.

Cornfield, J., Gordon, T., and Smith, W. W. (1961), Quantal Response Curves for Experimentally Uncontrolled Variables, *Bulletin de l'Institut International de Statistique*, **38**, Part 3, 97–115.

Cox, D. R. (1970), *The Analysis of Binary Data*, London: Methuen.

Efron, B. (1975), The Efficiency of Logistic Regression Compared to Normal Discriminant Analysis, *Journal of the American Statistical Association*, **70**, 892–898.

Finney, D. J. (1964), *Statistical Method in Biological Assay*, 2nd ed., London: Charles Griffin.

Fisher, R. A. (1936), The Use of Multiple Measurements in Taxonomic Problems, *Annals of Eugenics*, **7**, 179–188.

Gilbert, E. S. (1968), On Discrimination Using Qualitative Variables, *Journal of the American Statistical Association*, **63**, 1399–1412.

Goldberger, A. S. (1964), *Econometric Theory*, New York: Wiley.

Gordon, T. (1974), Hazards in the Use of the Logistic Function with Special Reference to Data from Prospective Cardiovascular Studies, *Journal of Chronic Diseases*, **27**, 97–102.

Halperin, M., Blackwelder, W. C., and Verter, J. I. (1971), Estimation of the Multivariate Logistic Risk Function: A Comparison of the Discriminant Function and Maximum Likelihood Approaches, *Journal of Chronic Diseases*, **24**, 125–158.

Hanushek, E. A., and Jackson, J. E. (1977), *Statistical Methods for Social Scientists*, New York: Academic Press.

Hauck, W. W. (1975), Some Aspects of the Logit Analysis of Quantal Response Bioassay, Ph.D. thesis, Harvard University.

Hauck, W. W., and Donner, A. (1977), Wald's Test as Applied to Hypotheses in Logit Analysis, *Journal of the American Statistical Association*, **72**, 851–853.

Hoel, P. G. (1962), *Introduction to Mathematical Statistics*, 3rd ed., New York: Wiley.

Jones, R. H. (1975), Probability Estimation Using a Multinomial Logistic Function, *Journal of Statistical Computation and Simulation*, **3**, 315–329.

Kahn, H. A., Herman, J. B., Medalie, J. H., Neufeld, H. N., Riss, E., and Goldbourt, U. (1971),

Factors Related to Diabetes Incidence: A Multivariate Analysis of Two Years Observation on 10,000 Men, *Journal of Chronic Diseases,* **23,** 617–629.

Mantel, N. (1973), Synthetic Retrospective Studies and Related Topics, *Biometrics,* **29,** 479–486.

Moore, D. H., II (1973), Evaluation of Five Discrimination Procedures for Binary Variables, *Journal of the American Statistical Association,* **68,** 399–404.

Mosteller, F., and Tukey, J. W. (1977), *Data Analysis and Regression: A Second Course in Statistics,* Reading, MA: Addison-Wesley.

Seigel, D. G., and Greenhouse, S. W. (1973), Multiple Relative Risk Functions in Case-Control Studies, *American Journal of Epidemiology,* **97,** 324–331.

Smith, M. S. (1973), Some Short-Term Effects of Project Head Start: A Report of the Third Year of Planned Variation—1971–1972, Cambridge, MA: Huron Institute.

Snedecor, G. W., and Cochran, W. G. (1967), *Statistical Methods,* 6th ed., Ames, IA: Iowa State University Press.

Stopher, P. R. (1969), A Probability Model of Travel Mode Choice for the Work Journey, *Highway Research Record,* **283,** 57–65.

Talvitie, A. (1972), Comparison of Probabilistic Modal-Choice Models: Estimation Methods and System Inputs, *Highway Research Record,* **392,** 111–120.

Theil, H. (1970), On the Estimation of Relationships Involving Qualitative Variables, *American Journal of Sociology,* **76,** 103–154.

Truett, J., Cornfield, J., and Kannel, W. (1967), A Multivariate Analysis of the Risk of Coronary Heart Disease in Framingham, *Journal of Chronic Diseases,* **20,** 511–524.

Wald, A. (1943), Tests of Statistical Hypotheses Concerning Several Parameters when the Number of Observations Is Large, *Transactions of the American Mathematical Society,* **54,** 426–482.

Walker, S. H., and Duncan, D. B. (1967), Estimation of the Probability of an Event as a Function of Several Independent Variables, *Biometrika,* **54,** 167–179.

Weisberg, H. I. (1973), Short-Term Cognitive Effects of Head Start Programs: A Report of the Third Year of Planned Variation—1971–1972, Cambridge, MA: Huron Institute.

Welch, B. L. (1939), Note on Discriminant Functions, *Biometrika,* **31,** 218–220.

CHAPTER 10

Log-Linear Analysis

Log-linear analysis can be used when the risk, outcome, and confounding variables are all categorical so that the data may be presented in a multidimensional table of counts called a *contingency table*. The dimension of the table is equal to the total number of variables (including outcome), and the entry in a given cell of the table is the count of individuals possessing a specified level or value of each variable. The following two examples of contingency tables will be used throughout the chapter to illustrate the basic features of log-linear analysis.

Example 10.1   UGDP data: The University Group Diabetes Program (UGDP) conducted a randomized trial to study the effectiveness and safety of different treatments for diabetes (this study was discussed briefly in Section 4.3). We concentrate here on two of those treatments—tolbutamide and a placebo—and relate them to one of the outcome variables measured by the UGDP investigators, cardiovascular mortality. Table 10.1 comes from a report on the UGDP findings (UGDP, 1970): it is a two-dimensional

Table 10.1   UGDP Data

Treatment	Outcome Cardiovascular Death	Other[a]	Total
Tolbutamide	26	178	204
Placebo	10	195	205
	36	373	409

[a] "Other" includes death from noncardiovascular causes and survivor.

Adapted, by permission of the American Diabetes Association, Inc., from UGDP (1970), Table 1.

or two-way table, since the individuals are classified along two variables (treatment and outcome). The treatment has two levels (placebo and tolbutamide) and the outcome also (cardiovascular death and other).

In the upper left-hand cell, for example, 26 indicates that there were 26 patients in the tolbutamide group who died due to a cardiovascular cause. The total number of patients under each treatment was fixed by the design: 204 and 205, respectively.

Example 10.2   Head Start data: In this cohort study (Smith, 1973; Weisberg, 1973), preschool children were divided into two groups, one receiving the Head Start program and the other not, and they were subsequently tested to assess the effect of the program. For each child the result of the test (pass/fail) was recorded along with program received and education of the head of household.

Table 10.2 presents the records of 491 children in a three-way contingency table, where the education of the head of household has three levels (less than 9 years of school, 10 and 11, or more than 12). There are, for instance, 28 children in the Head Start Program who failed the test and whose mothers had more than 12 years in school.

Table 10.2   Head Start Data

Education of the Head of Household	Outcome Test	Treatment Head Start	Control
9⁻	Fail	16	119
	Pass	1	31
10–11	Fail	23	83
	Pass	13	27
12⁺	Fail	28	86
	Pass	23	41

Data from Head Start Program (Smith, 1973; Weisberg, 1973).

Log-linear analysis can be viewed as a tool for understanding the structure of a contingency table such as Table 10.2 and more specifically for estimating treatment effects via the odds ratio. Simple tools for analyzing a two-way table such as Table 10.1 are the chi-square test for independence and the calculation of simple odds ratios. We will show that these tools can be expressed in terms of log-linear models and, in the two-dimensional case with dichotomous variables, log-linear analysis will tell us no more than these two techniques will tell us. However, for three and more dimensions, log-linear models allow us to go beyond the simple model of independence and explore other kinds of relationships between the variables. Indeed, we are concerned with the effect on cell counts not only of each variable taken independently, as in the model of independence, but also of interactions among all the variables—risk, outcome, confounding—and we want to single out the relevant interactions to assess the treatment effect.

The log-linear formulation is particularly appropriate in studies, case-control and cohort, where we want to measure the treatment effect by the odds ratio because, as we will show, there is a simple relationship between the odds ratio and the parameters of the log-linear model. Log-linear models derive their name from expressing the *log*arithm of expected cell counts as a *linear* function of parameters; these parameters represent the variables and their interactions, properly defined below. The use of logarithms will be justified below.

The maximum number of parameters one can use in the model is equal to the number of cells in the table. When we fit a log-linear model to a set of data, we try to represent the data with a minimal number of parameters, just as in regression we often represent the relationship between two variables with just an intercept and a slope.

We will first review the chi-square test for independence and the crude odds ratio and branch from there to log-linear models for two-way tables (Section 10.2). The more interesting log-linear models for three and more variables will be explored next (Section 10.3 and 10.4) and various aspects of log-linear models will be discussed (Sections 10.5 to 10.8).

This chapter will introduce the basic features of log-linear models. A detailed account of log-linear analysis may be found in Bishop et al. (1975). Recently published introductory books are Everitt (1977), Fienberg (1977), and Reynolds (1977). On a much higher mathematical level, there is Haberman's treatment of log-linear models (1974).

10.1 LOG-LINEAR MODELS FOR A TWO-DIMENSIONAL TABLE

We will introduce log-linear models with the simple 2 × 2 table from the UGDP study (Table 10.1) and spend some time on this very simple example in order to justify the use of the log transformation, to introduce notation and to explicate the connection with the odds ratio and the χ^2 test for independence.

The complexity of the log-linear approach will certainly appear disproportionate to the 2 × 2 situation, but the reader should remember that this is just a step toward understanding log-linear models for three and more variables, including confounding variables.

Suppose that we want to use the UGDP two-dimensional table to assess whether tolbutamide has an effect on cardiovascular mortality. A familiar approach is to use the chi-square test for independence. The idea underlying the chi-square test for independence is to assume that tolbutamide has no effect on mortality, to estimate expected numbers of individuals in each cell under that assumption, and to compare them with the observed counts through a Pearson chi-square statistic. The magnitude of the chi-square statistic indicates how far we are from the model of independence.

We follow this process step by step for our example. If tolbutamide had no effect, the cardiovascular death rate would be the same in the placebo and tolbutamide groups, and we would estimate it by 36/409, the ratio of the total number of cardiovascular deaths to the total number of individuals. We then multiply the death rate by the number of individuals in each treatment group to get estimated expected numbers of deaths. Table 10.3 presents the estimated expected table of counts for the model of independence or no treatment effect.

Table 10.3   Estimated Expected Table for UGDP Data under Model of Independence

	Cardiovascular Deaths	Other	Total
Tolbutamide	$\frac{36}{409} \times 204 = 17.96$	$\frac{373}{409} \times 204 = 186.04$	204
Placebo	$\frac{36}{409} \times 205 = 18.04$	$\frac{373}{409} \times 205 = 186.96$	205
Total	36	373	409

The reader can check that the odds ratio, defined in Chapter 3 as the ratio of the product of the diagonal cells, is equal to 1 in Table 10.3, as it should be since we computed it assuming no treatment effect.

The next step is to compare the estimated expected counts in Table 10.3 with the observed counts in Table 10.1. We calculate the Pearson chi-square statistic by summing the following expression over all cells of the contingency table

$$\frac{(\text{observed cell count} - \text{estimated expected cell count})^2}{\text{estimated expected cell count}},$$

which results in

$$\chi^2 = \frac{(26 - 17.96)^2}{17.96} + \frac{(178 - 186.04)^2}{186.04} + \frac{(10 - 18.04)^2}{18.04} + \frac{(195 - 186.96)^2}{186.96}$$

$$= 7.88.$$

Reference to a chi-square table with 1 degree of freedom indicates that the fit of the model of independence is poor (p value $< .01$). Therefore, we need to find another model to describe more adequately the structure of the UGDP data (i.e., by dropping the hypothesis of no treatment effect).

Before turning to another model, we will show that the model of independence can be written as a log-linear model. We need first to introduce some notation. Let $\hat{m}_{ij}$ denote the estimated expected counts in Table 10.3, where the subscript i stands for the ith level of the outcome variable ($i = 1$ or 2) and j stands for the jth level of the treatment ($j = 1$ or 2). Let x_{ij} denote the corresponding observed counts in Table 10.1. For example, we have $\hat{m}_{12} = 18.04$ and $x_{12} = 10$.

We see in Table 10.3 that $\hat{m}_{ij}$ is computed as

$$\hat{m}_{ij} = \frac{x_{i+}x_{+j}}{x_{++}}, \tag{10.1}$$

where a $+$ stands for summation over the levels of the variable. For example, $x_{i+} = x_{i1} + x_{i2}$. Taking natural logarithms (denoted by log) on both sides, we transform the multiplicative representation into an additive one:

$$\log \hat{m}_{ij} = -\log x_{++} + \log x_{i+} + \log x_{+j},$$

where $\log \hat{m}_{ij}$ appears as the sum of a constant (i.e., a term which depends on neither i nor j), a term depending on i only and a term depending on j only. This suggests writing each expected log count as a sum of a constant parameter, a parameter depending on the ith level of the outcome and a parameter depending on the jth level of the risk variable:

$$\log m_{ij} = u + u_i^Y + u_j^R, \tag{10.2}$$

where the superscripts Y and R refer to the outcome and risk variables, respectively, and the subscripts i and j refer to the levels of these variables. When talking of these parameters, u, u_i^Y, and u_j^R, we will sometimes simply refer to them as the u terms.

We now establish the correspondence between the u terms and the $\log m_{ij}$'s. We define u as the grand mean of the expected log counts,

$$u = \frac{1}{4} \sum_{i=1}^{2} \sum_{j=1}^{2} \log m_{ij},$$

$u + u_i^Y$ as the mean of the expected log counts at level i of Y,

$$u + u_i^Y = \frac{1}{2} \sum_{j=1}^{2} \log m_{ij},$$

and $u + u_j^R$ as the mean of the expected log counts at level j of R,

$$u + u_j^R = \frac{1}{2} \sum_{i=1}^{2} \log m_{ij}.$$

Since u_i^Y and u_j^R represent deviations from the grand mean u,

$$\sum_{i=1}^{2} u_i^Y = 0 \quad \text{and} \quad \sum_{j=1}^{2} u_j^R = 0. \tag{10.3}$$

Readers familiar with analysis of variance will have recognized at this point notations and zero-sum constraints on the parameters in (10.3) similar to those used in analysis of variance. The parallelism with analysis of variance will continue as we define other parameters.

Equation 10.2 represents the model of independence or of no treatment effect as a log-linear model for a two-way table. As there might be some confusion in terminology, we need to emphasize that treatment effect here has the same meaning as in the remainder of this book, namely the effect of the treatment or risk factor on the outcome. On the other hand, we also sometimes call the u terms in (10.2) effects; in particular, u_j^R is an effect of the risk factor on the expected log count: it has nothing to do with our usual "treatment effect."

Because of the zero-sum constraints (10.3), the representation (10.2) for the 2×2 contingency table amounts to specifying three parameters, namely u, u_1^Y, and u_1^R. We found, via the Pearson chi-square statistics, that the model (10.2) of no treatment effect does not fit the data adequately. Therefore, we extend the log-linear model to estimate a nonzero treatment effect and we do that by adding a fourth parameter, u_{ij}^{YR}, which is a two-factor effect between outcome and risk factor measuring the association between Y and R. The log-linear model is now

$$\log m_{ij} = u + u_i^Y + u_j^R + u_{ij}^{YR} \tag{10.4}$$

with an additional set of zero-sum constraints

$$\sum_{i=1}^{2} u_{ij}^{YR} = 0 \quad \text{for } j = 1, 2 \quad \text{and} \quad \sum_{j=1}^{2} u_{ij}^{YR} = 0 \quad \text{for } i = 1, 2; \tag{10.5}$$

so that, for our 2×2 table, u_{12}^{YR}, u_{21}^{YR}, and u_{22}^{YR} are determined once u_{11}^{YR} is specified.

Model (10.4) with its zero-sum constraints (10.5 and 10.3) is called the *saturated model* because it contains the maximum number of independent parameters (i.e., 4: u, u_1^Y, u_1^R, and u_{11}^{YR}) permissible to describe a set of four cells. In the saturated model, the estimated expected log counts are equal to the observed log counts and the fit is perfect.

Since the model of independence, with fewer parameters, did not fit the data, we select, for this example, the saturated model as the appropriate log-linear model.

By computing an odds ratio, we can understand more clearly how u_{ij}^{YR} is related to the treatment effect. By definition the odds ratio is

$$\psi = \frac{m_{11}m_{22}}{m_{12}m_{21}},$$

or, equivalently,

$$\log \psi = \log m_{11} + \log m_{22} - \log m_{12} - \log m_{21}.$$

We can express $\log \psi$ as a function of the u terms of (10.4):

$$\log \psi = u_{11}^{YR} + u_{22}^{YR} - u_{12}^{YR} - u_{21}^{YR}.$$

Using the zero-sum constraints in (10.5), we get

$$\log \psi = 4u_{11}^{YR}. \tag{10.6}$$

The relationship (10.6) provides us with a clear interpretation of the two-factor effect u_{ij}^{YR}: setting the two-factor effect equal to zero is equivalent to setting the log odds ratio equal to zero or the odds ratio equal to 1, which brings us back to the no-treatment-effect or independence model.

Finally, since (10.4) is our model of choice for our example, we estimate the treatment effect via the odds ratio on the basis of (10.4). Since it is a saturated model, the estimated expected odds ratio is equal to the observed odds ratio:

$$\hat{\psi} = \frac{26 \times 195}{10 \times 178} = 2.85. \tag{10.7}$$

The UGDP example has helped us to introduce two log-linear models for a 2×2 table. The more complicated model (10.4) differs from the simple one (10.2) by a parameter which is proportional to the log-odds ratio, so that the simpler one corresponds to a no-treatment-effect model. In the two-way table, we have not learned from log-linear models anything more about the effect of the treatment than is told us by the test for independence and the crude odds ratio (10.7). With three variables we will discover the advantages of the log-linear formulation, since we will have models which are intermediate between the independence model and the saturated model.

10.2 LOG-LINEAR MODELS FOR A THREE-DIMENSIONAL TABLE

Let us consider now a three-dimensional or three-way table where, besides treatment and outcome factors, we also have a potential confounding factor such as education of the head of household in the Head Start data. We saw in Chapter 3 that it is necessary to control for a confounding factor which otherwise would distort the estimate of treatment effect. To estimate the treatment effect, as measured here by the odds ratio, we will fit log-linear models of different complexity as in the two-dimensional case, pick one that fits the data adequately, and base our estimate on this appropriate model.

In the process of fitting a model, we will check that the potential confounding factor is really a confounding factor and we will see whether the treatment effect, as measured by the odds ratio, can be assumed constant over all levels of the potential confounding factor or whether there is evidence of interaction (see Section 3.3). We discuss here only a statistical technique for selecting a confounding factor, keeping in mind that judgment and collaboration with scientists are important parts of this process (see Section 5.1).

We extend the saturated model (10.4) to three factors by considering three main effects—one for each factor—three two-factor effects, and by introducing a three-factor effect. We will explain below the meaning of these effects that again we will call u terms. The saturated log-linear model for a three-dimensional table expresses the expected cell log count log m_{ijk} as

$$\log m_{ijk} = u + u_i^Y + u_j^R + u_k^X + u_{ij}^{YR} + u_{ik}^{YX} + u_{jk}^{RX} + u_{ijk}^{YRX}, \quad (10.8)$$

where the superscripts Y, R, and X refer to the outcome, risk, and confounding factors, respectively, and the subscripts i, j, and k refer to the levels of these three variables. Denote by I, J, and K the limits of these subscripts. For instance, in the Head Start data:

$$\begin{cases} i \text{ runs from 1 to 2 (fail/pass):} & I = 2 \\ j \text{ runs from 1 to 2 (Head Start/control):} & J = 2 \\ k \text{ runs from 1 to 3 } (9^-/10 - 11/12^+): & K = 3 \end{cases}$$

Again, we have zero-sum constraints on the parameters that are generalizations of (10.3) and (10.5):

$$\sum_{i=1}^{I} u_i^Y = \sum_{j=1}^{J} u_j^R = \sum_{k=1}^{K} u_k^X = \sum_{i=1}^{I} u_{ij}^{YR} = \sum_{j=1}^{J} u_{ij}^{YR} = \sum_{i=1}^{I} u_{ik}^{YX} = \sum_{k=1}^{K} u_{ik}^{YX}$$

$$= \sum_{j=1}^{J} u_{jk}^{RX} = \sum_{k=1}^{K} u_{jk}^{RX} = \sum_{i=1}^{I} u_{ijk}^{YRX} = \sum_{j=1}^{J} u_{ijk}^{YRX} = \sum_{k=1}^{K} u_{ijk}^{YRX} = 0. \quad (10.9)$$

It is necessary to impose these relationships to avoid an excess of parameters—remember that the number of independent parameters must be no more than the number of cells in the table. With these zero-sum constraints, the u terms appear as deviations from u terms of lower order.

All the log-linear models for three-dimensional tables are special cases of the saturated model (10.8), with one or more parameters set equal to zero. Again the connection between odds ratios and the parameters of the model helps us in interpreting the different models.

In a three-way table where $I = J = 2$, we can define an odds ratio for each level k of the potential confounding factor X. Say that

$$\psi_k = \frac{m_{11k}m_{22k}}{m_{12k}m_{21k}} \tag{10.10}$$

or, equivalently,

$$\log \psi_k = \log m_{11k} + \log m_{22k} - \log m_{12k} - \log m_{21k}.$$

With the log-linear model (10.8) we can express $\log \psi_k$ as a function of the u terms. We get

$$\log \psi_k = u_{11}^{YR} + u_{22}^{YR} - u_{12}^{YR} - u_{21}^{YR} + u_{11k}^{YRX} + u_{22k}^{YRX} - u_{12k}^{YRX} - u_{21k}^{YRX}.$$

Using the zero-sum constraints (10.9) to simplify the formula, we get

$$\log \psi_k = 4u_{11}^{YR} + 4u_{11k}^{YRX}. \tag{10.11}$$

Log ψ_k depends on the level k of the confounding factor X only through the three-factor effect u_{11k}^{YRX}. The only way we can have a constant odds ratio (i.e., no interaction, see Section 3.3) is with the three-factor effect equal to zero. Thus, if a log-linear model with no three-factor effect fits the data adequately, the data are not inconsistent with a hypothesis of constant odds ratio. We can then estimate this constant odds ratio by $4\hat{u}_{11}^{YR}$. The log-linear model with no three-factor effect is represented as

$$\log m_{ijk} = u + u_i^Y + u_j^R + u_k^X + u_{ij}^{YR} + u_{ik}^{YX} + u_{jk}^{RX}. \tag{10.12}$$

We will discuss in Section 10.5 the estimation of expected cell counts and thereby expected u terms in log-linear models. Let us just note at this point that the log-linear model (10.12) does not admit simple expressions for the estimated cell counts as did, for instance, the model of independence in the two-way table [see (10.1)].

The next level of simplicity is to assume one of the two factor effects equal to zero, keeping the three-factor effect equal to zero. Let the two-factor effect u_{ij}^{YR} relating risk to outcome factor be equal to zero for all (i, j). (We express this more concisely by $u^{YR} = 0$.) The model is then

$$\log m_{ijk} = u + u_i^Y + u_j^R + u_k^X + u_{ik}^{YX} + u_{jk}^{RX}.$$

From the formula for the log odds ratio (10.11), we see that setting $u^{YR} = u^{YRX} = 0$ amounts to assuming the odds ratio equal to 1 for each level of the confounding factor (i.e., independence between risk and outcome factors conditional on the confounding factor X). In such a model, we estimate the expected cell counts by

$$\hat{m}_{ijk} = \frac{x_{i+k}x_{+jk}}{x_{++k}}.$$

In general, a two-factor effect set equal to zero means independence of the two factors for a fixed level of the third one.

If we set two two-factor effects equal to zero, as in the model

$$\log m_{ijk} = u + u_i^Y + u_j^R + u_k^X + u_{ij}^{YR},$$

where both u^{YX} and u^{RX} are zero, then the omitted factor (here X) is jointly independent of the two others.

Note that when we defined a confounding variable in Section 2.1, we required it to be associated both with the outcome and risk factors. In the log-linear framework, this means that u^{YX} and u^{RX} measuring the association of X with the two other variables should both be nonzero if X is really a confounding factor.

Going a step further, we can set all three two-factor effects equal to zero:

$$\log m_{ijk} = u + u_i^Y + u_j^R + u_k^X. \tag{10.13}$$

This corresponds to a complete independence of the three factors. The estimated expected cell counts for such a model are familiar:

$$\hat{m}_{ijk} = \frac{x_{i++}x_{+j+}x_{++k}}{(x_{+++})^2}.$$

We have discovered in this section log-linear models intermediate between the simple model of independence between the three factors (10.13) and the saturated model (10.8) which fits the observed data perfectly. Interest in these intermediate log-linear models lies in particular in their ability to represent specific hypotheses about the treatment effect measured by the odds ratio.

10.3 LOG-LINEAR MODELS FOR MULTIDIMENSIONAL TABLES

Log-linear models give us a systematic way of analyzing contingency tables of any dimension. They can be defined for any number of factors. Let P be this number. We can in particular use log-linear models on a P-dimensional table with one outcome factor Y, one risk factor R, and $P - 2$ confounding factors X_1, X_2 and so on. The saturated model is an extension of (10.8) containing up to a P-factor effect, and the other log-linear models are special cases of the saturated model with one or more parameters set equal to zero.

As in the two- and three-dimensional tables, we want to focus on the parameters involving the risk and outcome factors in order to estimate the treatment effect. We will get a single measure of treatment effect, that is, an odds ratio constant over all levels of the confounding variables if all the three-factor and higher-order effects involving Y and R are equal to zero:

$$u^{YRX} = 0 \qquad \text{for all } X \text{ in } (X_1, X_2, \ldots, X_{P-2})$$

$$u^{YRXX'} = 0 \qquad \text{for all } (X, X') \text{ in } (X_1, X_2, \ldots, X_{P-2}) \qquad \text{etc.}$$

The parameters not involving Y, the outcome variable, reflect association between the background variables (i.e., risk variable and confounding variables). The u terms of order larger than three reflect associations which are difficult to interpret; therefore, we often try to fit only models with low-order u terms.

10.4   FITTING A LOG-LINEAR MODEL

In order to fit a log-linear model to a set of data, we need to be able to perform three tasks:

- We have to estimate the expected cell counts or the expected u terms under a postulated log-linear model.
- We have to assess the adequacy of the postulated model.
- We have to choose a satisfactory model.

When we analyzed the simple two-way table of the UGDP study in Section 10.2, we followed exactly that process. We assumed the model of independence, estimated the expected cell counts, rejected the model on the basis of the Pearson chi-square statistic and finally selected the saturated model as the only appropriate log-linear model for this table.

We discuss here the three steps involved in fitting a model and close this section by illustrating them with the analysis of the Head Start data.

10.4.1   Estimation

We restrict our attention to maximum likelihood estimation. For a given log-linear model, the estimates of the expected cell counts are the same whether we are dealing with a cohort or case-control study. The only concern related to the type of studies is to include in the model the parameters corresponding to the totals fixed by the design. In a case-control study the total number of cases and the total number of controls are fixed; in a cohort study, the numbers of individuals in the different treatment groups are fixed. For instance, in the UGDP data, an adequate log-linear model has to include u_j^R because the total number of patients x_{+j} in each treatment group was fixed by the design. The derivation of this rule is beyond the scope of this chapter, but it can be viewed as a consequence of a more general rule which says that the u terms included in a log-linear model determine the constraints on the estimated expected counts. For instance, model (10.2) contains u, u_i^Y, and u_j^R, so that the estimated expected counts satisfy the constraints

$$\hat{m}_{++} = x_{++} \qquad \text{because of } u$$

$$\hat{m}_{i+} = x_{i+} \qquad \text{because of } u_i^Y \qquad\qquad (10.14)$$

$$\hat{m}_{+j} = x_{+j} \qquad \text{because of } u_j^R.$$

For the UGDP study, once the x_{+j} were fixed by the design, we want $\hat{m}_{+j} = x_{+j}$. This is accomplished by putting u_j^R in the model.

The estimates of the expected counts satisfy those marginal constraints imposed by the u terms included in the model and satisfy the conditions imposed by the u terms set equal to zero. For instance, $u^{YR} = 0$ in model (10.2) implies that

$$\log \left(\frac{\hat{m}_{11}\hat{m}_{22}}{\hat{m}_{12}\hat{m}_{21}} \right) = 0. \qquad\qquad (10.15)$$

Remember that for model (10.2), we have

$$\hat{m}_{ij} = \frac{x_{i+}x_{+j}}{x_{++}}. \qquad\qquad (10.16)$$

The reader can check that these $\hat{m}_{ij}$ satisfy (10.14) and (10.15).

The estimates of the expected counts cannot always be expressed as an explicit function of observed marginal totals as was done in (10.16). Consider again the model (10.12) with no three-factor effect for a three-dimensional table. The constraints imposed by the u terms in the model reduce to

$$\hat{m}_{ij+} = x_{ij+}$$

$$\hat{m}_{i+k} = x_{i+k}$$

$$\hat{m}_{+jk} = x_{+jk}.$$

We have, moreover, $\hat{u}^{YRX} = 0$. The $\hat{m}_{ijk}$ solutions of this set of equations do not have an explicit form. In this case we use an iterative process to estimate the expected counts. Computer programs for handling log-linear models are now commonly available. Besides carrying on the iterative process for computing the estimated expected cell counts, they also provide estimated expected u terms (which allows us to estimate odds ratios) with their estimated asymptotic variances, goodness-of-fit statistics that we will discuss shortly, and other quantities of interest.

If the postulated model is true, and if the sample size is large, then the estimated u terms and in particular the estimated logarithm of the odds ratio will be approximately unbiased, since estimation is based on maximum likelihood. We do not know of any study of the bias of the estimated u terms in log-linear analysis for small samples.

10.4.2   Assessing Goodness of Fit

To assess goodness of fit of a postulated model, we follow the usual procedure of estimating expected values assuming that the model is true and comparing them with observed values. We compare them using either the Pearson chi-square statistic or the likelihood ratio statistic. Remember that the Pearson chi-square statistic is equal to the sum over all cells of

$$\frac{(\text{observed cell count} - \text{estimated expected cell count})^2}{\text{estimated expected cell count}}.$$

The likelihood ratio statistic is the sum over all cells of

$$2(\text{observed cell count}) \log \left(\frac{\text{observed cell count}}{\text{estimated expected cell count}} \right).$$

These statistics both have an asymptotic chi-square distribution with degrees of freedom given by

Degrees of freedom =
number of cells in the table—number of independent parameters.

Lack of fit of a model is indicated by large values of these statistics corresponding to small p values.

10.4.3   Choosing a Model

Selecting an appropriate log-linear model may be quite tedious if the dimension of the table is four or more, since many possible models are available. Different strategies based on stepwise methods which look at estimated expected u terms and goodness-of-fit statistics have been proposed and compared via some examples (Bishop et al., 1975, Chap. 4; Goodman, 1971).

The likelihood ratio statistic is particularly useful for comparing estimated expected counts under two different models where one model (the "simple" model) is a special case of the other one (the "complicated" model): some of the u terms of the complicated model are set equal to zero in the simple model (the models are said to be nested). Then the likelihood ratio statistic for the simple model can be expressed as the sum of two terms: one is the likelihood ratio statistic for the complicated model, the other is a conditional likelihood ratio statistic testing how far the estimated expected counts under the two models are from each other. This conditional statistic also follows asymptotically a chi-square distribution with degrees of freedom equal to the difference of degrees of freedom between the two models. If this chi-square value is small, the extra u terms in the complicated model are judged unimportant. Therefore, this approach that we will illustrate below allows us to judge the significance of u terms conditional on a complicated model.

We now turn to fitting a log-linear model to the Head Start data.

Example 10.2 **(continued):** The *saturated model* fits the data perfectly and, under that model, we estimate the treatment effect by the observed odds ratios, one for each level of the potential confounding factor, education of the head of household. We get

$$\hat{\psi}_{9-} = \frac{16 \times 31}{1 \times 119} = 4.17$$

$$\hat{\psi}_{10-11} = \frac{23 \times 27}{13 \times 83} = 0.58$$

$$\hat{\psi}_{12+} = \frac{28 \times 41}{23 \times 86} = 0.58.$$

Note that the odds ratio for 9 years or less of education is based on a table containing a 1 in one cell. This makes the estimate, 4.17, quite imprecise.

To simplify the model, we try the *model with no three-way effect*. Remember that we do not have direct estimates of the expected cell counts under that model, but must use an iterative procedure, usually performed by computer. This gives the estimated expected cell counts that appear in Table 10.4. The reader can check that the counts $\hat{m}_{ijk}$ in Table 10.4 satisfy the constraints imposed by the u terms in the model. The two goodness-of-fit statistics are:

$$\text{Pearson } \chi^2 = 4.20$$

$$\text{Likelihood ratio } \chi^2 = 5.17,$$

with 2 degrees of freedom each. Since the p value is greater than 0.075, we judge the fit of the model with no three-factor effect as adequate. Under this model, the treatment effect is constant over all levels of the head of household's education. We estimate the

Table 10.4   Head Start Example: Fitted Data for Model with $u^{YRX} = 0$

Education of the Head of Household	Test	Head Start	Control
9^-	Fail	12.89	122.11
	Pass	4.11	27.89
10–11	Fail	24.28	81.72
	Pass	11.72	28.28
12^+	Fail	29.83	84.17
	Pass	21.17	42.83

constant odds ratio either directly from the fitted table or from the estimated u terms. From the first level of education of the head of household we have, for instance,

$$\hat{\psi} = \frac{12.89 \times 27.89}{4.11 \times 122.11} = 0.72.$$

The reader should find the same number (except for rounding errors) from the other two levels. We get $\hat{u}_{11}^{YR} = -0.08$ and from $\log \hat{\psi} = 4u_{11}^{YR}$ [see (10.11)], we have

$$\log \hat{\psi} = 4 \times (-0.08) = -0.32 \Rightarrow \hat{\psi} = 0.73.$$

We can now check whether education of the head of household is really a confounding factor, by trying successively the *model with no outcome by confounding effect* ($u^{YX} = 0$, still keeping $u^{YRX} = 0$) and the *model with no risk by confounding effect* ($u^{RX} = 0$, with $u^{YRX} = 0$). The estimated expected cell counts under each model are given in Tables 10.5 and 10.6 together with the goodness-of-fit statistics.

Table 10.5   Head Start Example: Fitted Data for Model with
$$u^{YRX} = u^{YX} = 0$$

Education of the Head of Household	Test	Head Start	Control
9^-	Fail	10.95	111.63
	Pass	6.05	38.37
10–11	Fail	23.19	81.86
	Pass	12.81	28.14
12^+	Fail	32.86	94.51
	Pass	18.14	32.49

Pearson $\chi^2 = 13.52$
Likelihood ratio $\chi^2 = 15.39$ $\Big\}$ 4 degrees of freedom; p value $< .01$

From the chi-squares and their associated p-values, we judge that neither of these models fits the data. We conclude that education of the head of household is a confounding factor.

Let us now use Table 10.5 to illustrate the conditional likelihood ratio statistic. We choose the model with no three-factor effect ($u^{YRX} = 0$) as the complicated model; the likelihood ratio statistic under that model is 5.17, with 2 degrees of freedom. The simple model is the model with no outcome by confounding effect ($u^{YRX} = u^{YX} = 0$); the likelihood ratio statistic under that model is 15.39, with 4 degrees of freedom. Therefore, the conditional likelihood ratio statistic is $15.39 - 5.17 = 10.22$, with $4 - 2 = 2$ degrees of freedom. A p value smaller than 0.01 associated with 10.22 leads us to judge that u^{YX} is an important effect which should not be omitted from the model.

Finally, note that if we fit the *model with no outcome by risk effect,* we get the estimated expected cell counts in Table 10.7, which are acceptable based on the chi-square statistics. Under that model, the treatment effect, measured by the odds ratio, is equal

Table 10.6   *Head Start Example: Fitted Data for Model with*
$$u^{YRX} = u^{RX} = 0$$

Education of the Head of Household	Test	Head Start	Control
9^-	Fail	25.48	109.52
	Pass	8.71	23.29
10–11	Fail	20.01	85.99
	Pass	10.88	29.12
12^+	Fail	21.52	92.49
	Pass	17.41	46.59

Pearson $\chi^2 = 19.71$
Likelihood ratio $\chi^2 = 23.91$ } 4 degrees of freedom; p value $< .01$

to 1. We prefer to keep the model with no three-factor effect (Table 10.4) which yields the estimate of the odds ratio $\hat{\psi} = 0.73$, but this $\hat{\psi}$ is not significantly different from 1, since the model of Table 10.7 fits the data adequately. The computer program we used for this example did not give standard deviations on the u terms.

Table 10.7   *Head Start Example: Fitted Data for Model with*
$$u^{YRX} = u^{YR} = 0$$

Education of the Head of Household	Test	Head Start	Control
9^-	Fail	13.74	121.26
	Pass	3.26	28.74
10–11	Fail	26.14	79.86
	Pass	9.86	30.14
12^+	Fail	32.66	81.34
	Pass	18.34	45.66

Pearson $\chi^2 = 6.57$
Likelihood ratio $\chi^2 = 7.03$ } 3 degrees of freedom; p value $> .07$

This log-linear analysis of the Head Start data is not meant to be a realistic analysis of the Head Start program, since it represents a small subsample of the larger data set and since many other data on this experiment are available.

However, we hope it has served the purpose of illustrating how log-linear models can be used.

10.5   ORDERED CATEGORIES

The log-linear models we have considered so far treat all the factors as unordered categorical and do not exploit any information from possible ordering of categories of some of the factors. For instance, in the Head Start data, there is a natural ordering of categories of the confounding factor, education of the head of household, which we may want to include in the analysis. Special log-linear models have been developed for handling these situations. To use them, we must assign scores to the categories of the ordered factor(s). Fienberg (1977) gives a simple and clear presentation of these models.

10.6   RELATIONSHIP WITH LOGIT ANALYSIS ON CATEGORICAL VARIABLES

We will show in this section that logit analysis on categorical variables is a special case of log-linear analysis. Suppose that we have a three-dimensional contingency table with an outcome factor, a risk factor, and a confounding factor. Then a logit model with two independent variables (risk and confounding) can be derived from a log-linear model on the three-dimensional table. Suppose, as was the case in our examples, that both the outcome and risk factors are dichotomous and consider, for instance, the log-linear model with all the u terms included except the three-factor effect u_{ijk}^{YRX}:

$$\log m_{ijk} = u + u_i^Y + u_j^R + u_k^X + u_{ij}^{YR} + u_{ik}^{YX} + u_{jk}^{RX}. \qquad (10.17)$$

Remember from Chapter 9 that the logit of a probability p is defined as

$$\text{logit } p = \log \left(\frac{p}{1-p} \right).$$

Let us call "success" the first category of the outcome variable and "failure" the second. Then the probability of success p_{jk} in the jth category of the risk factor and the kth category of the confounding factor is

$$p_{jk} = \frac{m_{1jk}}{m_{1jk} + m_{2jk}}$$

$$\text{logit } p_{jk} = \log \left(\frac{p_{jk}}{1 - p_{jk}} \right) = \log \left(\frac{m_{1jk}}{m_{2jk}} \right).$$

Using the zero-sum constraints

$$u_1^Y + u_2^Y = 0$$
$$u_{1j}^{YR} + u_{2j}^{YR} = 0$$
$$u_{1k}^{YX} + u_{2k}^{YX} = 0,$$

we get

$$\text{logit } p_{jk} = 2u_1^Y + 2u_{1j}^{YR} + 2u_{1k}^{YX}. \tag{10.18}$$

Since the index of Y is constant in (10.18) (i.e., equal to 1), we can reparametrize this relationship by

$$w = 2u_1^Y$$
$$w_j^R = 2u_{1j}^{YR}$$
$$w_k^X = 2u_{1k}^{YX}.$$

Therefore,

$$\text{logit } p_{jk} = w + w_j^R + w_k^X \tag{10.19}$$

and the zero-sum constraints on the u terms impose

$$\sum_{j=1}^{2} w_j^R = 0 \quad \text{and} \quad \sum_{k=1}^{K} w_k^X = 0.$$

The correspondence between the logit model (10.19) and the logit model (9.7) can be easily established by treating R as a dummy variable and X as $K - 1$ dummy variables $X_1, \ldots, X_{K-1}$:

$$R = \begin{cases} 1 & \text{for category 1 of the risk factor} \\ 0 & \text{for category 2 of the risk factor} \end{cases}$$

$$X_1 = \begin{cases} 1 & \text{for category 1 of the condounding factor} \\ 0 & \text{for any other category of the confounding factor} \end{cases}$$

$$\vdots$$

$$X_{K-1} = \begin{cases} 1 & \text{for category } K - 1 \text{ of the confounding factor} \\ 0 & \text{for any other category of the confounding factor.} \end{cases}$$

(This method of creating $K - 1$ dummy variables to represent a K-category variable was presented in Section 9.8.)

The model (9.7) is

$$\text{logit } p(R, X_1, \ldots, X_{K-1}) = \alpha + \gamma R + \sum_{k=1}^{K-1} \beta_k X_k, \tag{10.20}$$

which amounts to

$$\text{logit } p_{1k} = \alpha + \gamma + \beta_k \quad \text{and} \quad \text{logit } p_{2k} = \alpha + \beta_k \qquad \text{for } k = 1, \ldots, k - 1$$
$$\text{logit } p_{1K} = \alpha + \gamma \qquad \text{and} \quad \text{logit } p_{2K} = \alpha$$

$$(10.21)$$

Equating the right-hand sides of (10.19) and (10.21), we obtain

$$\alpha = w + w_2^R + w_K^X$$

$$\alpha + \gamma = w + w_1^R + w_K^X$$

$$\alpha + \gamma + \beta_k = w + w_1^R + w_k^X \qquad k = 1, \ldots, K - 1.$$

Solving for the parameters in the logit model (10.20) yields

$$\alpha = w + w_2^R + w_K^X = 2u_1^Y + 2u_{12}^{YR} + 2u_{1K}^{YX}$$

$$\beta_k = w_k^X - w_K^X = 2u_{1k}^{YX} - 2u_{1K}^{YX} \qquad k = 1, \ldots, K - 1$$

and using the zero-sum constraints and (10.11),

$$\gamma = w_1^R - w_2^R = 2w_1^R = 4u_{11}^{YR} = \log \psi.$$

We now have a logit model with the logit parameters expressed in terms of the u terms of a log-linear model. To understand why logit analysis is a special case of log-linear analysis when all variables are categorical, note that in the derivation of the logit model passing from (10.17) to (10.18), all u terms involving only R and X dropped out. Therefore, other log-linear models than (10.17) lead to the same logit model (10.18). For example, one could set $u^{RX} = 0$ in (10.17):

$$\log m_{ijk} = u + u_i^Y + u_j^R + u_k^X + u_{ij}^{YR} + u_{ik}^{YX}. \qquad (10.22)$$

Fitting the log-linear models (10.17) and (10.22) would yield different estimated u terms and hence different estimates of γ, the logarithm of the odds ratio. If γ were to be estimated by the maximum likelihood procedure of Chapter 9, we would get the same estimate as that based on the log-linear model (10.17). So logit analysis implies a log-linear model with the u_{jk}^{RX} terms included. This is so because logit analysis by considering only the distribution of Y given the observed values of X and R is therefore implicitly treating all the marginal totals corresponding to the background variables as fixed (see Appendix 9A.1).

In large contingency tables in particular, we may want to fit log-linear models not including all the u terms involving all the backgound variables. By using a simpler model, we will reduce the variance of the estimated expected cell counts and u terms and, if the fit of the model is good, we can hope that their bias is small. Using a logit model approach does not give us this choice since it implicitly includes all the u terms involving all the background variables. The advantage of logit analysis, on the other hand, is that numerical variables can be included in the model without being forced to stratify them. A more extensive discussion of the comparison between log-linear and logit models can be found in Bishop (1969). Additional remarks appear at the end of Section 9.3.

10.7 OTHER USES OF LOG-LINEAR MODELS

We have emphasized the use of log-linear models for estimating treatment effects, but we may want to fit log-linear models to contingency tables for other purposes:

- To check whether a suspected confounding factor is really a confounding factor, as we did in Example 10.2. In particular, when there are many potential confounding factors, log-linear analysis of the multidimensional table in conjunction with knowledge of the area under investigation (see Section 5.1) provides a way to selecting the most important confounding factors. Bishop (1971) gives conditions for and implications of condensing the original table, (i.e., reducing its dimension by dropping potential confounding factors).
- To smoothe the observed data: whenever we are dealing with multidimensional tables, we may have very sparse data with large counts in a few cells and small ones elsewhere. One may improve the validity of a count in a particular cell by borrowing strength from neighboring cells, since they may contain information relevant to this particular cell (Mosteller, 1968). In particular, log-linear analysis provides a way of obtaining nonzero counts for cells where zeros are observed. Such a smoothing can be performed as a first step before standardization (Bishop and Mosteller, 1969), as was mentioned in Section 7.4.2.

REFERENCES

Bishop, Y. M. M. (1969), Full Contingency Tables, Logits, and Split Contingency Tables, *Biometrics,* **25,** 383–389.

Bishop, Y. M. M. (1971), Effects of Collapsing Multidimensional Contingency Tables, *Biometrics,* **27,** 545–562.

Bishop, Y. M. M., and Mosteller, F. (1969), Smoothed Contingency Table Analysis, *in* J. P. Bunker et al., Eds., *The National Halothane Study,* Washington, DC: U.S. Government Printing Office.

Bishop, Y. M. M., Fienberg, S. E., and Holland, P. W. (1975), *Discrete Multivariate Analysis— Theory and Practice,* Cambridge, MA: MIT Press.

Everitt, B. S. (1977), *The Analysis of Contingency Tables,* London: Chapman & Hall.

Fienberg, S. E. (1977), *The Analysis of Cross-Classified Categorical Data,* Cambridge, MA: MIT Press.

Goodman, L. A. (1971), The Analysis of Multidimensional Contingency Tables: Stepwise Procedures and Direct Estimation Methods for Building Models for Multiple Classifications, *Technometrics,* **13,** 33–61.

Haberman, S. J. (1974), *The Analysis of Frequency Data,* Chicago: University of Chicago Press.

Mosteller, F. (1968), Association and Estimation in Contingency Tables, *Journal of the American Statistical Association,* **63,** 1–28.

Reynolds, H. T. (1977), *The Analysis of Cross-Classifications,* New York: Free Press.

Smith, M. S. (1973), Some Short-Term Effects of Project Head Start: A Report of the Third Year of Planned Variation—1971–1972, Cambridge, MA: Huron Institute.

University Group Diabetes Program (1970), A Study of the Effects of Hypoglycemic Agents on Vascular Complications in Patients with Adult-Onset Diabetes, *Diabetes,* **19**(Suppl. 2), 787–830.

Weisberg, H. I. (1973), Short-Term Cognitive Effects of Head Start Programs: A Report on the Third Year of Planned Variation—1971–1972, Cambridge, MA: Huron Institute.

CHAPTER 11

Survival Analysis

In many longitudinal studies the outcome variable is the time elapsed between the entry of a subject into the study and the occurrence of an event thought to be related to the treatment. The following are examples of such studies that will be referred to throughout this chapter.

Example 11.1 Leukemia remission times: Freireich et al. (1963) compare the times in remission of a group of leukemia patients treated with the drug 6-mercaptopurine, and of an untreated control group. The event of interest is the first relapse, that is, the end of the remission period. The outcome variable is the time from entry into the study (when all subjects were in remission) to the first relapse.

Example 11.2 Stanford heart transplant program: Patients who are judged to be suitable recipients of a transplanted heart enter a queue and receive a heart if and when a donor is found. The outcome variable is the elapsed time between the date the patient was judged suitable and the event of interest, death. The survival of patients who receive a heart is to be compared with that of patients who do not.

Example 11.3 Duration of unemployment: Lancaster (1978) considers the effect of various risk factors such as age, ratio of unemployment benefit received to wage received in last job, and proportion of the labor force out of work, on the length of time unemployed. The outcome variable here is the time between the date of first interview and of getting a job.

To achieve a uniform terminology we may refer to the event of interest as "death" and the outcome variable as "survival time," recognizing that in many applications, such as Example 11.3, the event may not be death and may even be a desirable occurrence.

Longitudinal studies may have subjects entering at various dates throughout the period of the study. For example, some subjects may enter a study lasting from January 1979 to January 1983 at its inception, others may enter at various times between these two dates. The date of entry of each subject into the study must be recorded since survival time for that subject is usually calculated from that date. In this chapter, unless otherwise stated, time means time from entry to the study.

A serious complication in the analysis of survival times is the possibility of censoring, that is, of subjects not being observed for the full period until the occurrence of the event of interest.

This may happen for several reasons:

1. The study may be terminated while some subjects are still alive.
2. If deaths from a specific cause are of interest, the subject may die of another cause, thought to be unrelated. For example, in Example 11.2 it may be thought that the death of a study subject from a road accident should not be counted as evidence for or against transplantation, yet it effectively terminates the study as far as that subject is concerned.
3. The subject may be lost to follow up during the course of a study through relocating or changing treatment regime.

Section 11.1 discusses the simplest form of analysis of survival data—using the concept of "total time at risk"—and the rather restrictive assumptions under which this method is valid. Section 11.2 shows how a life table, constructed from censored data, may be used to estimate the distribution of survival times in a homogeneous group of subjects. Methods of comparing life tables are discussed in Section 11.3, which also introduces the important *proportional hazards model*. In Section 11.4 we show how an extended proportional hazards model can provide estimates of treatment effect adjusted for confounding factors. The life-table estimate of the distribution of survival times can also be adjusted for confounding factors. This is discussed in Section 11.5. Analytical and graphical tests of the proportional hazards model are introduced in Section 11.6, after a short discussion of time-dependent background variables.

The choice of appropriate starting point for the measurement of survival is as important as the choice of end point. In assessing the effect of a treatment on survival, it is natural to take as the starting point the date when the treatment was actually received. However, this can lead to biased assessments if there is an appreciable chance that a subject will die before the treatment is administered. Section 11.7 illustrates a special technique for handling studies such as the Stanford heart transplant program (Example 11.2) where there is a delay before treatment is applied.

Serious problems of interpretation can arise, when, as in reasons 2 and 3 given above, the censoring may be related to the event of interest. Section 11.8 discusses these issues and briefly describes the competing risks model for the analysis of deaths from several causes. Section 11.9 mentions some alternative approaches to the analysis of survival data.

In this chapter, as in the entire book, we focus on comparative studies which aim to assess the effect on survival of a specified treatment or risk factor. Our methods are well suited also to other kinds of study. Life tables (Section 11.2) provide useful summaries of the survival experience of single groups of subjects. This descriptive aspect has been extensively developed in actuarial work and in population studies, which use a rather different notation and terminology from that of Section 11.2. Cox's regression model (Section 11.4, Appendix 11C) is a valuable technique in exploratory studies which try to sift through many different background variables for those which may influence survival. (As always, the results of such "fishing expeditions" need to be interpreted with great care.)

The appendixes give a little more mathematical detail and some formulas. Appendix 11A gives precise definitions of the terms "hazard function" and "survivor function," which are informally defined in Section 11.2, and states the relationships betwen the two functions in discrete and in continuous time. A slightly improved form of the life-table estimator of the distribution of survival times is given in Appendix 11B. Cox's regression model, which underlies the proportional hazards models of Sections 11.3 and 11.4, is specified in fuller generality in Appendix 11C. We indicate how parameter estimates, standard errors, and significance tests may be obtained. Finally, Appendix 11D describes Breslow's estimator of the survivor function introduced in Section 11.5.

Peto et al. (1976, 1977) include a nontechnical account of life tables and the log-rank test in a general discussion of the design and analysis of clinical trials. Texts such as those of Bradford Hill (1977), Colton (1974), and Armitage (1971) also describe medical applications of life tables. A pioneering paper by Cox (1972) is the source for many of the methods discussed in Sections 11.4 to 11.8. Two substantial texts, each requiring more mathematical background than we assume, are Gross and Clark (1975) and Kalbfleisch and Prentice (1979).

11.1   THE TOTAL TIME AT RISK

The total time at risk gives a simple procedure for comparing survival times in two groups subject to censoring. After describing the calculations, we discuss two assumptions necessary for the validity of the method: constancy of risk over time and homogeneity of the groups. Example 11.1 (comparison of remission times of leukemia patients treated with the drug 6-mercaptopurine and of an untreated control group) will illustrate the technique.

Table 11.1 gives the number of weeks each patient was under observation. The asterisks denote incomplete follow-up. For example, the entry 6* in the drug group represents a subject who was still in remission after 6 weeks and was not observed after that time. But the unasterisked 6 refers to a subject known to be in remission at 6 weeks but to have relapsed in the seventh week. Note that "relapse" here is the outcome of interest, corresponding to "death" in the terminology introduced in the introduction to this chapter, and survival times are here interpreted as times in remission.

Although it is immediately clear from Table 11.1 that the treatment group have generally longer remission times than the control group, it is not obvious

Table 11.1   Remission Times of Leukemia Patients[a]

[Treatment group: 6*, 6, 6, 6, 7, 9*, 10*, 10, 11*, 13, 16, 17*, 19*, 20*, 22, 23, 25*, 32*, 32*, 34*, 35*]

[Control group: 1, 1, 2, 2, 3, 4, 4, 5, 5, 8, 8, 8, 8, 11, 11, 12, 12, 15, 17, 22, 23]

[a] An asterisk indicates incomplete follow-up.

how a simple quantitative measure of the superiority of the treatment group can be derived. We could attempt to compare *average remission times*. Unfortunately, the censored remission times would bias such a comparison, whether they are included in the calculations or not. If the censored times are excluded, the treatment group average will be too low, as it will not include the longest remission times (25*, 32*, etc.). An average calculated ignoring the distinction between deaths and censored survival times would also be unfair to the treatment group, since most of the actual deaths occurred in the control group.

Another approach often used in clinical studies is to calculate *survival rates*. Crude survival rates represent the proportion of patients who survive a specified length of time, for example, 5 years or, more appropriately in Example 11.1, 10 weeks. Censoring again causes difficulties. The leukemia patient in the treatment group of Table 11.1 who was lost to follow-up after 20 weeks must be counted as a "survivor" in a calculation of a 10-week survival rate, but the subject who was lost to follow-up after 6 weeks could not be so counted, since although he or she was known to have survived 6 weeks, he or she may not have survived for the full 10 weeks.

If all censoring takes place at the end of a study, the time when each subject would be censored if he or she had not previously died would be known. Unbiased estimates of 5-year (say) survival rates can be obtained from those subjects who entered the study at least 5 years before its termination. For none of these subjects will have been, nor could have been, censored. Subjects who entered the study less than 5 years before its termination must be excluded from the calculation, even if they die, since they could not possibly have been observed to survive for the full 5-year period. Although the exclusion of the recent patients will avoid bias, it will also lose precision. Moreover, no crude survival rate can be calculated if some censoring occurs during the course of a study, as the appropriate denominator, the number of subjects who might have been observed to die before being censored, is unknown. For data subject to censoring we do not recommend the use of crude survival rates, preferring the life-table estimates to be discussed in Section 11.2.

Returning to Example 11.1, a quite different approach is to estimate, for each group, the risk of death (relapse) of a "typical" patient during a "typical" week. We first calculate the total time at risk for each group as the sum of the censoring and survival (remission) times for that group. The average risk of death (relapse) for each group is calculated as the number of deaths (relapses) in that group divided by the corresponding total time at risk.

The treatment and control groups have total times at risk of

$$6 + 6 + \cdots + 35 = 359$$

$$1 + 1 + \cdots + 23 = 182,$$

respectively. These are the numbers of person-weeks in which a relapse could have occurred. There were 9 relapses in the treatment group and 21 in the control group. The estimated risks for the two groups are

$$\frac{9}{359} = 0.025 \quad \text{and} \quad \frac{21}{182} = 0.115,$$

respectively. Note that censored observations contribute to the denominators but not to the numerators of these risk calculations.

The ratio of the two risks, $0.025/0.115 = 0.22$, is analogous to the relative risk discussed in Chapter 3. It indicates that the risk of relapse for a treated subject in any week is only 0.22 of the risk for a control subject.

11.1.1   Assumptions

The risk calculations just described attempt to estimate a probability of relapse (death in general) per person-week of observation in each group. For the estimates to make sense, in each group the probability of death must be the same for all person-weeks of observation. In the fanciful language of coin tossing, we assume that the survival experience of the two groups is determined by two biased coins, with probabilities λ_1 and λ_0 of showing heads (death) on any toss. Each week independent tosses of the appropriate coin are made for each subject who survives to the start of that week. If the coin shows heads, the subject dies during that week; if it shows tails, he or she survives until the next week. [The calculated risks for each group (here 0.025 and 0.115) are estimates of these probabilities λ_1 and λ_0.] Our assumption that these parameters do not vary can be split into two parts.

Assumption 1: Risk of death is independent of time. The chance of death of a subject in any week does not depend on how many weeks he or she has survived up to the start of that week.

Assumption 2: Homogeneity of groups. All subjects in each group have the same distribution of survival time.

Assumption 1 would be violated, for example, in comparative studies of surgical procedures where the risk of death during or immediately following an operation may be much higher than at a later time. A simple comparison of risks as presented above will give too much weight to the possibly very few long survival times, and too little weight to the early deaths.

The existence of measured or unmeasured background factors influencing survival will violate Assumption 2. Any such factor that is also related to the treatment will be a confounding factor and result in biased estimates of treatment effect. Section 11.4 discusses the derivation of estimates of treatment effect corrected for measured confounding factors.

11.2 LIFE TABLES

Life tables can describe the distribution of survival times experienced by homogeneous groups of subjects, when these survival times are subject to censoring. The risk of death need not be independent of time. Thus assumption 2 of Section 11.1 must hold; assumption 1 need not. This section will discuss the construction and interpretation of a life table for a single group. We shall use the data of Example 11.1, calculating the life table for both groups but concentrating most of our attention on the treatment group.

Notice that all remission and censoring times were reported as whole numbers of weeks. This has led to some ties in the data, which could have been resolved had the times been reported to a greater accuracy. The presence of such ties is a minor nuisance in survivorship data, and our discussion will include a simple correction procedure. We shall assume that the data were "truncated" rather than "rounded." That is, a reported time to censoring or relapse of 6 weeks is taken to mean an actual time of between 6 and 7 weeks, rather than of between $5\frac{1}{2}$ and $6\frac{1}{2}$ weeks.

To describe the distribution in the absence of censoring of the remission times of the treatment group of leukemia patients of Example 11.1, we would like to determine the *survivor function* $S(t)$. This function specifies, for each value of t, the probability that the remission time of a typical patient in the group would last beyond time t if there were no censoring. For example, $S(10)$ is the probability that the remission of a typical patient lasts 10 weeks or more. If there were no censoring the value $S(t)$ could be estimated, for each $t = 1, 2, \ldots$, by the survival rate at time t, that is, by the proportion of patients who survive for at least t weeks. In the control group, where there was no censoring, 8 patients out of a total of 21 have remissions lasting 10 weeks or more, so that $S(10)$ would be estimated by $8/21 = 0.3810$. Unfortunately, as was shown in Section 11.1, this simple estimate cannot be used if there is censoring of the data.

The life table is a method for estimating the survivor function $S(t)$ from censored data. Let us first define the time-dependent risk of death or *hazard function* $\lambda(t)$ to be the (conditional) probability that a patient who is known to be in remission at least until time t will relapse before time $t + 1$. In the actuarial literature, $\lambda(t)$ is called the force of mortality; in other contexts it is sometimes known as the age-specific failure rate. The $\lambda(t)$ can be estimated from censored data. Consider the patients who are known to be in remission at time t, because they have neither relapsed nor been censored before time t. Let $r(t)$ be the number of such patients, the number "at risk" between times t and $t +$ 1. A patient whose reported censoring time is t, and who therefore was actually censored at some time between t and $t + 1$, is counted as being at risk for half of the interval. Thus, in the treatment group of Example 11.1, $r(7) = 17$, since 17 patients had remissions lasting 7 weeks or more and no subjects were censored in the eighth week. But $r(6) = 20\frac{1}{2}$, since the patient reported as censored after 6 weeks is counted as being at risk for half of the seventh week. Let $m(t)$ denote the number of patients who relapse between t and $t + 1$ weeks. Then $m(6) = 3, m(7) = 1$. The hazard, $\lambda(t)$, is estimated by the proportion of relapses among those patients at risk between t and $t + 1$ weeks. That is,

$$\hat{\lambda}(t) = \frac{m(t)}{r(t)},$$

and in the example

$$\hat{\lambda}(6) = \frac{3}{20.5} = 0.1463$$

and

$$\hat{\lambda}(7) = \frac{1}{17} = 0.0588.$$

Tables 11.2 and 11.3 present complete calculations of $\hat{\lambda}(t)$ for the treatment and control groups of Example 11.1. However, our main interest is not in $\hat{\lambda}(t)$ itself, but in estimates for the survival probabilities $S(t)$ which may be derived from it. To see the relationship between $\lambda(t)$ and $S(t)$, consider a large number n of patients in a situation where there is no censoring. In the first week, about $n[\lambda(0)]$ patients will relapse, since the expected proportion of relapses is $\lambda(0)$ and in the first week, n patients are at risk. At the end of the first week, $n - n[\lambda(0)] = n[1 - \lambda(0)]$ patients will still be in remission, so

$$S(1) = \frac{n[1 - \lambda(0)]}{n} = 1 - \lambda(0).$$

At the start of the second week there are $n[1 - \lambda(0)]$ patients at risk. The expected proportion of relapses in the second week is $\lambda(1)$, and the expected number of relapses is $n[1 - \lambda(0)]\lambda(1)$. At the end of the second week, the expected number of patients still in remission is $n[1 - \lambda(0)] - n[1 - \lambda(0)]\lambda(1)$ $= n[1 - \lambda(0)] [1 - \lambda(1)]$. So the expected proportion of the original n patients

Table 11.2   Life Table for Treatment Group

Time, t	At Risk, $r(t)$	Relapses, $m(t)$	Hazard $\hat{\lambda}(t)$	$1 - \hat{\lambda}(t)$	$\hat{S}(t+1)$
0	21	0	0	1	1
1	21	0	0	1	1
2	21	0	0	1	1
3	21	0	0	1	1
4	21	0	0	1	1
5	21	0	0	1	1
6	20½	3	0.1463	0.8537	0.8537
7	17	1	0.0588	0.9412	0.8034
8	16	0	0	1	0.8034
9	15½	0	0	1	0.8034
10	14½	1	0.0690	0.9310	0.7480
11	12½	0	0	1	0.7480
12	12	0	0	1	0.7480
13	12	1	0.0833	0.9167	0.6857
14	11	0	0	1	0.6857
15	11	0	0	1	0.6857
16	11	1	0.0909	0.9091	0.6234
17	9½	0	0	1	0.6234
18	9	0	0	1	0.6234
19	8½	0	0	1	0.6234
20	7½	0	0	1	0.6234
21	7	0	0	1	0.6234
22	7	1	0.1429	0.8571	0.5343
23	6	1	0.1667	0.8333	0.4452
24	5	0	0	1	0.4452
25	4½	0	0	1	0.4452
26	4	0	0	1	0.4452
27	4	0	0	1	0.4452
28	4	0	0	1	0.4452
29	4	0	0	1	0.4452
30	4	0	0	1	0.4452
31	4	0	0	1	0.4452
32	4	0	0	1	0.4452
33	2	0	0	1	0.4452
34	1½	0	0	1	0.4452
35	½	0	0	1	0.4452

who are still in remission at the end of the second week is

$$S(2) = \frac{n[1 - \lambda(0)][1 - \lambda(1)]}{n} = [1 - \lambda(0)][1 - \lambda(\dot{1})].$$

This reasoning may be extended to show that for any time t,

$$S(t + 1) = [1 - \lambda(0)][1 - \lambda(1)] \cdots [1 - \lambda(t)].$$

Table 11.3   Life Table for Control Group

Time, t	At Risk, $r(t)$	Relapses, $m(t)$	Hazard $\hat{\lambda}(t)$	$1 - \hat{\lambda}(t)$	$\hat{S}(t + 1)$
0	21	0	0	1	1
1	21	2	0.0952	0.9048	0.9048
2	19	2	0.1053	0.8947	0.8095
3	17	1	0.0588	0.9412	0.7619
4	16	2	0.1250	0.8750	0.6667
5	14	2	0.1429	0.8571	0.5714
6	12	0	0	1	0.5714
7	12	0	0	1	0.5714
8	12	4	0.3333	0.6667	0.3810
9	8	0	0	1	0.3810
10	8	0	0	1	0.3810
11	8	2	0.2500	0.7500	0.2858
12	6	2	0.3333	0.6667	0.1905
13	4	0	0	1	0.1905
14	4	0	0	1	0.1905
15	4	1	0.2500	0.7500	0.1428
16	3	0	0	1	0.1428
17	3	1	0.3333	0.6667	0.0952
18	2	0	0	1	0.0952
19	2	0	0	1	0.0952
20	2	0	0	1	0.0952
21	2	0	0	1	0.0952
22	2	1	0.5000	0.5000	0.0476
23	1	1	1.0000	0	0

The corresponding combination of the values of the estimated hazard function $\hat{\lambda}(t)$ yields the *life-table* (or *actuarial*) *estimate* of the survivor function, as

$$\hat{S}(t + 1) = [1 - \hat{\lambda}(0)][1 - \hat{\lambda}(1)] \cdots [1 - \hat{\lambda}(t)].$$

Tables 11.2 and 11.3 also present the full calculations for the survivor function for the treatment and control groups of Example 11.1. Although it is only necessary to calculate $\hat{\lambda}(t)$ and $\hat{S}(t + 1)$ at times t where relapses occur, we have filled out the tables to emphasize the fact that implicit estimates of $\lambda(t)$ and $S(t + 1)$ are obtained for each time t up to the largest recorded relapse or censoring time. A nice feature of the calculation is that $\hat{S}(t + 1)$ may be found by multiplying $\hat{S}(t)$ by $1 - \hat{\lambda}(t)$. Figure 11.1 graphs the survivor functions for the two groups. In the control group, where there is no censoring, $\hat{S}(t)$ is just the proportion of the original 21 subjects who are still in remission at time t.

If the largest observation is a relapse (death, in general), the estimated value of the survivor function reaches zero at that time. However, if the largest observation is censored (as in Table 11.2), the estimated survivor function does

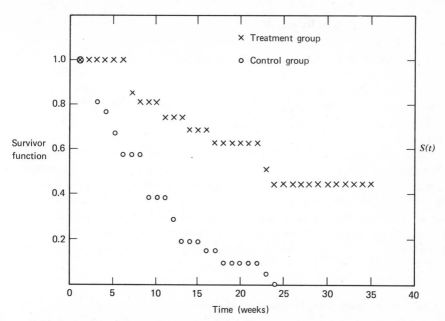

Figure 11.1   Life table estimates of the survivor functions for the treatment and control groups of Example 11.1.

not reach zero. It must be emphasized that the data provides no information concerning $S(t)$ for values of t beyond the range of the data.

11.2.1   Standard Errors

A rather complicated approximate formula for the standard error of $S(t)$ first given by Greenwood (1926) is often quoted. See, for example, Colton (1974, Chap. 9). Peto et al. (1977) suggest a much simpler approximation, in our notation

$$\text{Standard error } \hat{S}(t) = \hat{S}(t) \sqrt{\frac{1 - \hat{S}(t)}{r(t)}}.$$

This formula estimates the standard error of $\hat{S}(10)$ in the treatment group (Table 11.2) as

$$\text{Standard error } \hat{S}(10) = 0.8034 \sqrt{\frac{1 - 0.8034}{14.5}}$$

$$= 0.0935.$$

Peto et al. point out that this formula reduces to the usual expression for the standard error of a binomial proportion when there is no censoring. They state that the estimate is usually slightly too large, unlike the Greenwood formula, which may grossly underestimate the standard error for large values of t.

In any case, it should be noted that the values of $S(t)$ for large t are estimated with much less relative precision than are those for small t. This is because the life-table calculations for large t can be extremely sensitive to small changes in the data. For example, changing the censored remission time of 34* in Table 11.2 to a relapse at 34 weeks would change the estimated survivor function $\hat{S}(35)$ from 0.4452 to $(0.4452 \times \frac{1}{2}) = 0.2226$.

To derive approximate confidence limits for the value of the survivor function $S(t)$ at a single prespecified value of t, we can apply normal theory and use either Greenwood's or Peto's standard error formulas. However, this method cannot be used to give simultaneous confidence limits on several or on all values of $S(t)$. Approximate confidence bands on the entire function $S(t)$ can be derived, but this is beyond the scope of the book. Relevant references include Breslow and Crowley (1974), Meier (1975), Aalen (1976), and Gillespie and Fisher (1979).

11.2.2 Connection with the Constant Risk Model

We stated that life-table calculations require homogeneous groups [assumption 2 of Section 11.1] but not time-independent risks (assumption 1). If assumption 1 does hold, the true hazard function $\lambda(t)$ will not depend on time and we may write $\lambda(t) = \lambda$ for the risk of death, per unit of time, for each living subject in a homogeneous group. The survivor function for that group then takes the geometric form

$$S(t) = (1 - \lambda)^t. \tag{11.1}$$

The continuous-time version of (11.1) is the exponential

$$S(t) = e^{-\lambda t}, \tag{11.2}$$

where λ now denotes the hazard rate per unit time.

To see whether this model is a reasonable fit to the data, we can estimate λ as in Section 11.1 and compare the survivor function computed from (11.1) with the life-table estimate $S(t)$. Cox (1978, 1979) discusses this and other graphical methods for assessing the goodness of fit of the exponential function (11.2) to survival data. One useful technique, due to Nelson (1972), is to plot the ordered uncensored survival times t_i against the *estimated cumulative hazard*, defined for data with no tied survival times as

$$\hat{\Lambda}(t_i) = \hat{\lambda}(1) + \hat{\lambda}(2) + \cdots + \hat{\lambda}(t_i)$$

$$= \frac{1}{r(t_1)} + \frac{1}{r(t_2)} + \cdots + \frac{1}{r(t_i)},$$

since $\hat{\lambda}(t) = 1/r(t)$ if there is a death at t; $\hat{\lambda}(t) = 0$, otherwise. If the survival times follow the exponential distribution with hazard rate λ, $(\hat{\Lambda}(t_i), t_i)$ should fall close to a straight line of slope λ through the origin. Tied survival times should be plotted as if they were (just) distinct (i.e., by arbitrarily breaking the ties).

11.2.3   Effect of Grouping

The grouping intervals used to calculate the life table should be as narrow as is consistent with the accuracy of the data. Had the data of Example 11.1 been recorded daily rather than weekly, many of the reported ties between censoring and remission times would have been avoided and more accurate estimates obtained.

Kaplan and Meier (1958) show how an estimator of $S(t)$ [although not of $\lambda(t)$] may be obtained without any grouping of the survival times, provided that these have been recorded exactly. They call the resulting estimator the *product-limit estimator*, in contrast to the *actuarial estimator* discussed here. The two estimates will usually be very similar. Appendix 11B presents the formula for the product-limit estimator.

Computationally, the difference between the two estimators is that in computing the numbers at risk for the product-limit estimator, no "half-period" adjustment is made for the censored values. Thus in the product-limit estimator, $r(t)$ counts all subjects whose reported death or censoring time is t or greater.

11.3   COMPARISON OF LIFE TABLES

We can assess the effect of treatment on survival by comparing the life tables of the different treatment groups. Often a graphical presentation will suffice to indicate the direction and approximate magnitude of major differences between the groups. Figure 11.1 graphs the survival functions calculated in Tables 11.2 and 11.3 and clearly indicates the superiority of the treatment group.

Various measures of survival calculated from a life table can be used to compare the survival of groups. Summary measures often quoted in the literature include:

1. The median survival time.
2. The estimated survivor function at t, where the time t must be specified in advance of seeing the data.

Both these measures are easily calculated from the life table. The standard error for 2 is given in Section 11.2. In comparing the estimated survival functions $S(t)$ from two independent life tables, the standard error of the difference between the two estimates is computed in the usual way as the square root of the sum of squares of the standard errors of each estimate.

Unfortunately, these summary measures are generally inefficient since they use only part of the information in the life table. This section describes in detail the "log-rank" significance test for the comparison of two lifetables and mentions some alternative tests. We then introduce the proportional hazards model and show how this permits the estimation of the relative risk θ in a more general context than that of Section 11.1.

11.3.1   The Log-Rank Test

The idea behind the log-rank test for the comparison of two life tables is simple: if there were no difference between the groups, the total deaths occurring at any time should split between the two groups in approximately the ratio of the numbers at risk in the two groups at that time. So if the numbers at risk in the first and second groups in (say) the sixth year were 70 and 30, respectively, and 10 deaths occurred in that year, we would expect

$$10 \times \frac{70}{70 + 30} = 7$$

of these deaths to have occurred in the first group, and

$$10 \times \frac{30}{70 + 30} = 3$$

of the deaths to have occurred in the second group.

A similar calculation can be made at each time of death (in either group). By adding together for the first group the results of all such calculations, we obtain a single number, called the *extent of exposure* (E_1), which represents the "expected" number of deaths in that group if the two groups had the same distribution of survival time. An *extent of exposure* (E_2) can be obtained for the second group in the same way. Let O_1 and O_2 denote the actual total numbers of deaths in the two groups. A useful arithmetic check is that the total number of deaths $O_1 + O_2$ must equal the sum $E_1 + E_2$ of the extents of exposure.

The discrepancy between the O's and E's can be measured by the quantity

$$\chi^2 = \frac{(|O_1 - E_1| - \frac{1}{2})^2}{E_1} + \frac{(|O_2 - E_2| - \frac{1}{2})^2}{E_2}.$$

For rather obscure reasons, χ^2 is known as the log-rank statistic. An approximate significance test of the null hypothesis of identical distributions of survival time in the two groups is obtained by referring χ^2 to a chi-square distribution on 1 degree of freedom.

Example 11.1 (***continued***): Table 11.4 presents the calculations for the log-rank test applied to the remission times of Example 11.1. Note that the numbers at risk in each group and the number of deaths at each time come from the corresponding life tables, Tables 11.2 and 11.3. A chi-square of 13.6 is highly significant ($P < .001$), again indicating the superiority of the survival of the treatment group over that of the control group.

Table 11.4 Log-Rank Calculation for the Leukemia Data

Time, t	At Risk T	C	Total	Relapses T	C	Total	Extent of Exposure T	C	Total
1	21	21	42	0	2	2	1.0000	1.0000	2
2	21	19	40	0	2	2	1.0500	0.9500	2
3	21	17	38	0	1	1	0.5526	0.4474	1
4	21	16	37	0	2	2	1.1351	0.8649	2
5	21	14	35	0	2	2	1.2000	0.8000	2
6	$20\frac{1}{2}$	12	$32\frac{1}{2}$	3	0	3	1.8923	1.1077	3
7	17	12	29	1	0	1	0.5862	0.4138	1
8	16	12	28	0	4	4	2.2857	1.7143	4
10	$14\frac{1}{2}$	8	$22\frac{1}{2}$	1	0	1	0.6444	0.3556	1
11	$12\frac{1}{2}$	8	$20\frac{1}{2}$	0	2	2	1.2295	0.7705	2
12	12	6	18	0	2	2	1.3333	0.6667	2
13	12	4	16	1	0	1	0.7500	0.2500	1
15	11	4	15	0	1	1	0.7333	0.2667	1
16	11	3	14	1	0	1	0.7857	0.2143	1
17	$9\frac{1}{2}$	3	$12\frac{1}{2}$	0	1	1	0.7600	0.2400	1
22	7	2	9	1	1	2	1.5556	0.4444	2
23	6	1	7	1	1	2	1.7143	0.2857	2
Total				9	21	30	19.2080	10.7920	30
				(O_1)	(O_2)		(E_1)	(E_2)	

Illustration: $t = 23$, $2 \times \dfrac{6}{7} = 1.7143$, $2 \times \dfrac{1}{7} = 0.2857$.

Test of significance:

$$\chi^2 = \frac{\left(|O_1 - E_1| - \frac{1}{2}\right)^2}{E_1} + \frac{\left(|O_2 - E_2| - \frac{1}{2}\right)^2}{E_2}$$

$$= \frac{\left(|9 - 19.2| - \frac{1}{2}\right)^2}{19.2} + \frac{\left(|21 - 10.8| - \frac{1}{2}\right)^2}{10.8} = 13.6$$

Estimate of relative risk:

$$\hat{\theta} = \frac{9/19.2}{21/10.8} = 0.24$$

The log-rank test as presented by Peto et al. (1977) uses the product-limit life-table calculations (Section 11.2.3) rather than the actuarial estimators shown in Tables 11.2 and 11.3. As discussed in Appendix 11D, the distinction is unlikely to be of practical importance unless the grouping intervals are very coarse.

Peto and Pike (1973) suggest that the approximation in treating the null distribution of χ^2 as a chi-square is conservative, so that it will tend to understate the degree of statistical significance. Peto (1972) and Peto and Peto (1972) give a permutational "exact test," which, however, is applicable only to randomized studies. In the formula for χ^2 we have used the continuity correction of subtracting $\frac{1}{2}$ from $|O_1 - E_1|$ and $|O_2 - E_2|$ before squaring. This is recommended by Peto et al. (1977) when, as in nonrandomized studies, the permutational argument does not apply. Peto et al. (1977) give further details of the log-rank test and its extension to comparisons of more than two treatment groups and to tests that control for categorical confounding factors. The appropriate test statistics are analogous to the corresponding chi-squares for contingency tables (Snedecor and Cochran, 1967, Chap. 9).

11.3.2 Estimation: The Proportional Hazards Model

In the analysis of Example 11.1 by the total time at risk approach of Section 11.1, we were able to calculate an estimate $\hat{\theta}$ of the relative risk θ. Recall that θ is the ratio of the risk in the treatment group to that in the control group. When the hazard function $\lambda(t)$ for each group depends on time, the total time at risk approach is inappropriate. However, it may still be reasonable to assume that the hazard functions are proportional, that is, that, for each time t

$$\lambda_1(t) = \theta \lambda_0(t), \tag{11.3}$$

where the subscripts 1 and 0 refer to the treatment and control groups, respectively. This assumption implies that the effect of the treatment is to multiply the hazard at all times by the factor θ, which is assumed not to depend on time.

It can be shown that when (11.3) holds, there is an approximate relationship between the corresponding survivor functions of the form

$$S_1(t) = [S_0(t)]^\theta,$$

which may also be written as

$$\log [S_1(t)] = \theta \log [S_0(t)].$$

The approximation here is due to the discreteness of the data; the relation would be exact if survival times were measured exactly (in continuous time). A more precise definition of the $\lambda(t)$ as instantaneous rates of failure given in Appendix 11A avoids the need for approximation.

It is possible to derive various ad hoc procedures for the estimation of θ in (11.3), based on weighted and/or constrained linear regression applied to the estimated survivor functions from Tables 11.2 and 11.3. Cox (1972), who introduced the proportional hazards model, describes a better method for estimating θ, which is sketched in Appendix 11C. A computer will generally be required. A test of the null hypothesis that $\theta = 1$ is also available. Cox's method applied to the leukemia data of Example 11.1 yields $\hat{\theta} = 0.19$, here close to the estimate $\hat{\theta} = 0.22$ obtained from the "total time at risk" approach.

An alternative estimator suggested by Peto et al. (1977) uses the log-rank calculations. From Table 11.4 we form, for each group, the ratio of the observed number of deaths in each group to the total "extent of exposure to risk" in each group. The estimate of relative risk is the ratio of these two quantities, here $\theta = 0.24$. Cox's estimator is preferable to the log-rank estimator, although it requires more computation. Breslow (1975) indicates that the log-rank estimator of θ is biased toward unity, even in large samples. He states, however, that the bias is minimal for $\frac{1}{2} \leqslant \theta \leqslant 2$, only becoming noticeable for extreme values of θ.

Yet another approach is to apply the Mantel–Haenszel estimator (Sections 7.6 and 7B.3) for the common odds ratio in a series of 2×2 contingency tables. The interpretation of the treatment effect θ as an odds ratio rather than a relative risk is justified provided that the grouping intervals used to calculate the hazard functions $\lambda_0(t)$ and $\lambda_1(t)$ are sufficiently fine to ensure that these hazards are both small (see Chapter 3). In fact, the exact discrete-time proportional hazards model given in Appendix 11C defines θ to be an odds ratio.

Each row of Table 11.4 can be reexpressed as a 2×2 contingency table: for example, Table 11.5 gives the final row, corresponding to $t = 23$.

Table 11.5   Reexpression of the Final Row of Table 11.4

	Deaths	Survivors	Total
Treatment group	1 (a_i)	5 (b_i)	6
Control group	1 (c_i)	0 (d_i)	1
Total	2	5	7 (n_i)

The Mantel–Haenszel estimator of θ is computed by the formula

$$\hat{\theta} = \frac{\sum a_i d_i / n_i}{\sum b_i c_i / n_i},$$

where the symbols are as defined in Table 11.5. For these data we find that $\hat{\theta} = 0.194$, in good agreement with Cox's analysis. This calculation includes the "half-counts" corresponding to censored observations. Breslow (1975) quotes an unpublished remark of D. G. Clayton that the Mantel–Haenszel estimator closely approximates the maximum likelihood (Cox) estimator when θ is near unity.

Although the computational details of Cox's method are beyond the scope of the book, it is worth noting that they are very similar to those of the logistic analysis of matched samples in case-control studies (Section 13.3.2). A death at time t is regarded as a "case" whose matched "controls" are all subjects still alive and in the study at time t. Liddell et al. (1977) discuss this interpretation of the proportional hazards model in the context of a mortality study of asbestos workers.

11.3.3   Other Significance Tests

The log-rank test, apparently first proposed by Mantel (1966), has many competitors in the literature. Cox (1953) derived an approximate F test under the constant risk model. Gehan (1965a, b) proposed a version of the Wilcoxon two-sample test applicable to censored data. Mantel (1967) gave an alternative procedure for computing this statistic. Efron (1967) modified Gehan's test to achieve greater power and permit estimation of a parameter (not the relative risk θ) measuring the difference between the two survivor functions. Breslow (1970) extended Gehan's test to the comparison of more than two samples.

The choice of which test to use depends on whether the proportional hazards model is the likely alternative hypothesis. If so, then Cox's (1972) likelihood test or the log-rank test, which is closely related to it, should be used. According to Tarone and Ware (1977), who compared the two tests, the modified Wilcoxon statistic gives greater weight than the log-rank statistic to differences occurring near the beginning of a study and is less sensitive to events occurring when very few individuals remain alive.

11.4   INCLUSION OF BACKGROUND FACTORS

The straightforward comparison between the treatment and control groups discussed in Section 11.3 may be distorted by confounding factors. Fortunately, the proportional hazards model extends easily to include the effects of other factors besides the treatment of interest on the outcome. Cox's method can test whether a background factor, or a combination of several background factors, affects the outcome, and it provides estimates of the treatment effect adjusted for confounding factors. Although an important feature of this regression approach, like that of logit analysis with which it has much in common, is its ability to handle several background factors simultaneously, for ease of exposition we restrict ourselves to a single, numerical, confounding factor and a dichotomous treatment. Appendix 11C introduces the more general model.

The *extended proportional hazards model* assumes a risk of death at time t of

$$\lambda_1(t) = \theta\lambda^*(t) \exp (\beta_1 X) \tag{11.4}$$

for individuals in the treatment group and

$$\lambda_0(t) = \lambda^*(t) \exp (\beta_0 X) \tag{11.5}$$

for individuals in the control group. Here x denotes the value of the background variable, which differs from individual to individual, and β_1 and β_0 are coefficients which measure the influence of this background variable on the hazards in the two groups. If the two groups differ on X, then X will be a confounding variable. The parameter θ is the ratio of hazards for a treatment and control subject each with $X = 0$. The function $\lambda^*(t)$ is supposed unknown. Since (11.5) with $X = 0$ yields $\lambda_0(t) = \lambda^*(t)$, we can interpret $\lambda^*(t)$ as the hazard for a control subject with $X = 0$.

The ratio of hazards for a treatment and control subject each with the same nonzero value of X is $\theta \exp (\beta_1 X - \beta_0 X)$. This will vary with X unless $\beta_1 = \beta_0$ when its value will be θ irrespective of X. In the terminology of Chapter 3, $\beta_1 = \beta_0$ means no interaction between treatment and confounding variable, so that the effect of the treatment is to multiply the hazard by θ regardless of the level of the background variable.

If we set $\phi = \ln \theta$ ($\theta = e^\phi$), define the treatment indicator variable R by

$$R = \begin{cases} 0 & \text{for the control group} \\ 1 & \text{for the treatment group,} \end{cases}$$

and set $\delta = \beta_1 - \beta_0$, then (11.4) and (11.5) may be written as a single equation,

$$\lambda(t) = \lambda^*(t) \exp (\phi R + \beta_0 X + \delta X). \tag{11.6}$$

Note that the term δXR is zero for control subjects and is $(\beta_1 - \beta_0)X$ for treated subjects.

The method of partial likelihood, discussed briefly in Appendix 11C and more fully in Cox (1972, 1975) can be used to fit the model (11.6). Specifically, a likelihood ratio test of the hypothesis of no interaction ($\delta = 0$, or equivalently $\beta_1 = \beta_0$) is available. If no interaction is found, a likelihood ratio test of the hypothesis of no treatment effect ($\phi = 0$) may be performed. The partial likelihood also gives estimates, with standard errors, of all coefficients. In particular, $\hat{\theta} = e^{\hat{\phi}}$ is an estimate, adjusted for the background variable, of the true ratio of hazards θ.

Example 11.1 (*continued*): The initial white blood cell count is known to influence prognosis in leukemia. High white blood cell counts suggest short remission times. Table

11.6 presents hypothetical values for the common logarithm of the white blood cell count for the leukemia subjects of Example 11.1. (We use the logarithms to avoid giving undue weight to the very large counts.) These data raise the possibility that the tendency to longer remission times in the treated subjects may be due to their (on average) more favorable white blood cell counts.

Table 11.6　Hypothetical Log (White Blood Cell Counts) for the Leukemia Subjects of Table 11.1

Treatment Group		Control Group	
Survival Time[a]	log (WBCC)	Survival Time	log (WBCC)
6*	3.20	1	2.80
6	2.31	1	5.00
6	4.06	2	4.91
6	3.28	2	4.48
7	4.43	3	4.01
9*	2.80	4	4.36
10*	2.70	4	2.42
10	2.96	5	3.49
11*	2.60	5	3.97
13	2.88	8	3.52
16	3.60	8	3.05
17*	2.16	8	2.32
19*	2.05	8	3.26
20*	2.01	11	3.49
22	2.32	11	2.12
23	2.57	12	1.50
25*	1.78	12	3.06
32*	2.20	15	2.30
32*	2.53	17	2.95
34*	1.47	22	2.73
35*	1.45	23	1.97
(mean)	2.64	(mean)	3.22

[a] An asterisk indicates incomplete follow-up.

To test the hypothesis of no interaction, we compare twice the change in log likelihood when δ is included in the model to a chi-square on 1 degree of freedom (Appendix 11C). Here the interaction term is clearly nonsignificant ($\chi_1^2 = 0.43$). Maximum likelihood estimates of ϕ and β in the model

$$\lambda(t) = \lambda^*(t) \exp(\phi R + \beta X) \tag{11.7}$$

without the interaction term are

$$\hat{\phi} = -1.29 \quad \text{(standard error} = 0.422)$$
$$\hat{\beta} = 1.60 \quad \text{(standard error} = 0.329).$$

This analysis confirms that the white blood cell count is a strong indicator of remission duration, but also shows that difference in the white blood cell counts of the two groups cannot explain away the apparent treatment effect. The adjusted estimate of the ratio of hazards namely $\hat{\theta} = e^{-1.29} = 0.27$, is closer to unity than the unadjusted estimate $\hat{\theta}$ = 0.19 given in Section 11.3, but it is still highly significant ($\chi_1^2 = 10.39$).

11.5 ESTIMATING THE DISTRIBUTION OF SURVIVAL TIME

Although primary attention will usually focus on the estimator $\hat{\theta} = e^{\hat{\phi}}$ of treatment effect, the corresponding survivor functions $S_1(t)$ and $S_0(t)$ may also be of interest. In Section 11.2 these were estimated by separate life-table calculations for the treatment and control groups. With heterogeneous groups as discussed in Section 11.4, the life-table calculations are not appropriate, as they do not control for effect of the background factor. Even with homogeneous groups, we may wish to derive estimates of $S_1(t)$ and $S_0(t)$ which reflect and make use of the assumption (11.3) of proportional hazards.

The method proposed by Cox (1972) for estimating survivor functions within the context of the proportional hazard model has not been widely used. In the discussion published with Cox's paper, Oakes (1972) and Breslow (1972) proposed simpler approaches, which were later elaborated in Breslow (1974). Kalbfleisch and Prentice (1973) propose a slightly different procedure. The common idea behind all these methods is to approximate the underlying hazard function $\lambda^*(t)$ by a function that is constant over suitably defined grouping intervals of the time scale. The constant for each interval is estimated separately by maximum likelihood. Appendix 11D discusses Breslow's proposals in more detail.

11.6 TESTING THE PROPORTIONAL HAZARDS MODEL

For the life-table comparison of two homogeneous groups discussed in Section 11.3, we saw that the assumption of proportional hazards,

$$\lambda_1(t) = \theta\lambda_0(t), \tag{11.3}$$

is fundamental to the derivation of a single estimator of the realtive risk θ. Fortunately, this assumption can be tested in several ways and if the data suggest a particular form of departure from the assumption, a modified analysis can take this into account. Methods of testing for proportional hazards fall naturally into two types, analytical and graphical, and we consider each type in turn, after a brief discussion of time-dependent explanatory variables.

11.6.1   Time-Dependent Explanatory Variables

The discussion of the Cox model so far has assumed that although the treatment and confounding variables differ among individuals, for each individual each variable is constant over time (i.e., has a single value). This assumption can be relaxed by including explanatory variables that depend on time. For example, if blood pressure were thought to influence survival, the blood pressure of a patient on a particular day may be more relevant to the patient's risk of death on that day than the same patient's blood pressure on entry to the study. Each patient would have a series of blood pressure readings, the series terminating when the patient dies, or is censored.

Fortunately, Cox (1975) demonstrated that his method can cope with such time-dependent explanatory variables, provided that they also vary between individuals. Cox's method cannot assess time-dependent effects common to all patients, as these are subsumed into the unknown function $\lambda^*(t)$.

Section 11.7 gives an example where primary interest centers on the effect of a time-dependent treatment variable.

11.6.2   Analytical Tests for the Proportional Hazards Model

Here we show how the inclusion in the hazard function of a suitably chosen time-dependent background variable in the manner of Section 11.4 leads to a simple test of the assumption of proportional hazards. The new variable represents an interaction effect between the treatment indicator variable and time.

Example 11.1 (*continued*): Cox (1972) defined a new variable

$$z(t) = \begin{cases} t - 10 & \text{for the treatment group} \\ 0 & \text{for the control group.} \end{cases}$$

The hazard functions for the two groups become

$$\lambda_1(t) = \lambda^*(t) \exp\left[\phi + \beta(t - 10)\right]$$

$$\lambda_0(t) = \lambda^*(t),$$

where the coefficients ϕ and β are to be estimated from the data. The ratio of hazards in the two groups becomes

$$\frac{\lambda_1(t)}{\lambda_0(t)} = \exp\left[\phi + \beta(t - 10)\right],$$

which, according as β is positive or negative, will be an increasing or decreasing function of time. The constant 10, close to the mean survival time in the two groups, is inserted for computational convenience.

The estimate of the coefficient β given by Cox's procedure indicates the direction and approximate magnitude of any time trend in the ratio of hazards. A test of the hypothesis that $\beta = 0$ is also available. In this example no evidence of such a time trend was found.

If desired, other time-dependent terms, for example quadratic functions of time, can be fitted to search for more complicated time trends in the ratio of relative risks.

11.6.3 Graphical Methods

Graphical methods for testing the proportional hazards model (11.3) proposed by Kay (1977) and Crowley and Hu (1977) rely on the notion of generalized residuals discussed by Cox and Snell (1968). If the underlying hazard function $\lambda_0(t)$ and the coefficient θ were known, each time to death could be transformed to an exponentially distributed variable by a suitable reexpression of the time scale. Although the treatment and control groups will need separate formulas, the result would be a single series of independent exponentially distributed variables with the same mean. Censored times to death could similarly be transformed to censored exponential variables.

Although the function $\lambda(t)$ and coefficient θ are not generally konwn, they can be estimated by the methods of Sections 11.4 and 11.5, and reexpression formulas obtained using the estimates in place of the true values. A plot of the ordered values of the generalized residuals against their expected values under the exponential distribution gives some indication of the goodness of fit of the model. This approach is similar to the cumulative hazard plots mentioned in Section 11.2.

Unfortunately, the sampling theory of generalized residuals is complicated and has apparently not been investigated in survival analysis. It would seem that departures from the model would have to be quite substantial before they could be detected in this way.

11.6.4 Effect of Violations of the Model

If the proportional hazard model does not fit the data, calculation of a single estimate $\hat{\theta}$ of treatment effect is inappropriate, as the relative risk of the treatment will vary and must be estimated by a function of time rather than by a single value. In Example 11.2 (Section 11.6.2), had the coefficient $\hat{\beta}$ of $z(t)$ been significantly different from zero, we would have estimated the time-dependent relative risk as

$$\hat{\theta}(t) = \exp[\hat{\phi} + \hat{\beta}(t - 10)],$$

where $\hat{\phi}$ and $\hat{\beta}$ are the estimated coefficients of R and z given by Cox's analysis.

11.6.5 Extensions

Although this section has concentrated on testing the basic proportional hazards model (11.3), both the analytical and graphical methods generalize easily to the extended proportional hazards model introduced in Section 11.4 to handle confounding factors. Following the analytical approach, we may include in the fitted model terms representing interactions between confounding factors and time, as well as or instead of the term $z(t)$, which represents an interaction between the risk factor and time. The goodness of fit of the various models can be assessed by the likelihood ratio statistics, as described in Section 11.4 and Appendix 11C.

Or, following the methods of Section 11.6.3, we may derive reexpression formulas to convert the survival times into a series of independent and identically distributed exponential variables. These reexpression formulas will involve the treatment variable, the confounding variable, and their estimated coefficients. Kay (1977) and Crowley and Hu (1977) give examples of this technique.

11.7 ALLOWANCE FOR DELAY IN TREATMENT

Scarcity of resources may enforce a delay between the entry of a subject into the study and the application of a treatment. If this delay results in an appreciable number of deaths or losses to follow-up before the treatment is applied, estimates of treatment effect that do not take into account the delay or the deaths before treatment may be biased, for the subjects who actually receive a treatment will have been selected from the original study population by their ability to survive the delay. This is also a difficulty commonly encountered when evaluating intensive-care units or other emergency services.

In these situations the choice of starting point for the measurement of survival becomes unclear. Peto et al. (1977) suggest that in randomized studies, survival should be measured from the data of randomization and that for the purposes of statistical analysis, all subjects allocated to a particular treatment should be regarded as having received that treatment on the date of the allocation. A comparison of one treatment against another becomes in effect a comparison of a *policy* of applying the first treatment where possible against a *policy* of applying the second treatment where possible. Provided that appropriate statistical methods are used, as discussed in earlier sections, subsequent failure to administer the allocated treatment will not bias this comparison.

Although a similar approach can sometimes be used in nonrandomized studies, often these will have no decision point comparable to the date of randomization. Sometimes the only control or comparison group for the evaluation of a treatment will be those subjects who were intended to receive the same treatment but who did not receive it; naive comparisons can then be very misleading.

Example 11.2 (*continued*): The study population consists of all patients judged suitable for heart transplantation. Although the conference date when the potential heart recipient is selected provides a natural starting point for the study of subsequent survival time, patients cannot be operated on until a donor is found. Since many patients will die before a donor can be found, there are two categories of patients: those who receive a heart and those who do not. Their respective case histories are summarized in Figure 11.2. Some patients are still alive at the end of the study.

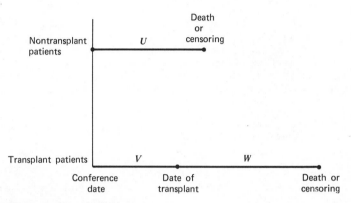

Figure 11.2 The heart transplant study.

We label the survival time (in days) of a typical nontransplant patient by U. For a transplant patient the time is labeled $V + W$, where V is the waiting time between the conference date and transplant date, and W is the survival time after the transplant operation. Note that the times to death in each group may be censored.

The first published evaluations compared the values of W and U. The mean values, $\overline{U} = 108$ and $\overline{W} = 709$, were computed using the "total time at risk" approach discussed in Section 11.1. (If patients alive at the end of the study are excluded, the calculations give $\overline{U} = 29.2$ and $\overline{W} = 203.8$.)

But there is a serious bias in this approach, as was pointed out by Gail (1972). The principal reason a transplant candidate does not receive a heart is that he or she dies before a donor is found. The patients who survive until transplantation might be hardier, and so likely to survive longer after transplantation, regardless of the merits of the operation.

Crowley and Hu (1977) give an ingenious alternative analysis which avoids this bias. They define a time-dependent treatment indicator variable R as follows:

$$R = \begin{cases} 0 & \text{if the candidate has not at that time received a new heart} \\ 1 & \text{if the candidate has at that time received a new heart.} \end{cases}$$

For a nontransplant patient, $R = 0$ throughout the period U of Figure 11.2 between the conference date and censoring, or death, whereas a transplant patient would have $R = 0$ for the period V between the conference date and transplant, and $R = 1$ for the period W between transplant and death. If no other background variables are considered, the hazard at time t after the conference date becomes

$$\lambda(t) = \lambda^*(t)$$

if the patient has not at that time received a heart and

$$\lambda(t) = e^{\beta}\lambda^*(t)$$

if the patient has by that time received a new heart. Transplantation will be beneficial if $\beta < 0$.

Crowley and Hu (1977) perform several analyses, including other important background variables, in particular, age. Their tentative conclusion is that "transplantation can prolong survival for certain younger patients if a suitably matched heart can be found."

Earlier evaluations of the Stanford heart transplant program along similar lines were given by Turnbull et al. (1974) and by Mantel and Byar (1974). The problem of delay between the decision to apply a treatment and its actual application can occur in many studies. The method of analysis described here requires that the amount of delay (time V of Figure 11.2) be recorded for each subject who does receive the treatment and also requires knowledge of the survival and censoring times of the untreated patients.

11.8   SELF-CENSORING: COMPETING RISKS

An important strength of the methods discussed in this chapter is that they do not require equal patterns of censoring among treatment groups. The validity of the analysis is not affected if the censoring is more severe in one treatment group than in another.

However, the analyses do require that the censoring not be related to the outcome. The prognosis for a censored subject at the time of censoring must not be affected by the knowledge that the subject is censored. In Example 11.1 (leukemia data), the analysis assumes that a patient in the treatment group who is censored at 10 weeks (say) would have the same expected (future) remission time as a patient in the treatment group who has remained in remission for 10 weeks and is *not* censored. In Example 11.2 (Stanford heart transplant program), the assumption is that a patient who is operated on (say) 10 weeks after the conference date is typical of all selected but not yet transplanted patients who are still alive 10 weeks after the conference date.

If the censoring pattern is related to the outcome, serious difficulties can arise. As an extreme example, consider the hospital that transfers all its terminally ill patients to another institution just before their death. If these are treated as censored observations, the hospital can truthfully, if meaninglessly, claim a zero death rate among its terminally ill patients!

A more realistic example is given by Fisher and Kanarek (1974). They quote a study concerning a medically supervised exercise program for people with cardiovascular problems, including those with a previous history of myocardial infarction. A subsequent follow-up study showed that those who drop out of the study are at an elevated risk for a heart attack, compared with those who stay in. Failure to allow for this censoring bias in the analysis of the study would lead to an overoptimistic assessment of the effectiveness of the program.

If the censoring is related to the outcome but not to the treatment, then Cox's model may lead to valid comparisons between groups even if the life-table estimate for a single group is biased. For example, suppose that in the Cox model (11.3) there is a probability $d(t)$ (which may depend on t) that a death at time t is recorded as a censoring. Although the true hazard rates are

$$\lambda_0(t) = \lambda^*(t)$$
$$\lambda_1(t) = \theta\lambda^*(t),$$

as before, the recorded hazard rates will be

$$\lambda_0'(t) = [1 - d(t)]\lambda^*(t)$$
$$\lambda_1'(t) = [1 - d(t)]\theta\lambda^*(t).$$

Cox's method applied to the recorded death rates will estimate the ratio of these two hazard functions. Since the term $[1 - d(t)]$ cancels, this ratio equals the ratio θ of the true hazard functions. Although the misrecording of deaths as censored values will reduce the precision of the estimate of θ, it will not introduce bias.

When the censoring is related to both the risk factor and the outcome, a biased and inconsistent estimator of θ will result. The only fully satisfactory solution is to obtain more data, perhaps by a further study, on the fate of the subjects who are lost to follow-up. If only a small proportion of the observations are censored, "optimistic" and "pessimistic" analyses for each treatment group may be constructed by arbitrarily setting subsequent survival after censoring to be infinite or zero, respectively. Comparison of the optimistic analysis for one group with the pessimistic analysis for another group will yield a lower limit to the estimated relative risk, and an upper limit can be found similarly.

11.8.1 Competing Risks

More generally, we may consider deaths from several different causes, including by convention, censoring as one "cause." A considerable literature has arisen on this topic, known as competing risks. [See, for example, Chiang (1961a, b; 1970) and Lagakos (1978).] If the different causes of death are unrelated (here we speak of independent risks), separate life-table estimates of the survival distribution corresponding to each cause of death can be obtained by applying the methods of Section 11.2, taking deaths from all causes other than the one under consideration as censored observations.

If, as often happens, the different causes of death are related, so that a subject with a higher-than-average risk of death from one cause is likely also to have a higher-than-average risk of death from other causes, the independent risk model may be inappropriate. For example, in a cancer study, a severely ill patient may commit suicide because of the cancer. This means that the distribution of cancer deaths that would have occurred among the suicides had they not committed suicide would differ from the distribution of cancer deaths among those who did not commit suicide. Regarding the suicides as censored observations would result in an underestimate of the true cancer death rate. Thus in cancer studies it may be safer to estimate treatment effects using deaths from all causes.

These issues have been discussed from a theoretical standpoint by Cox (1959), Gail (1975), Tsiatis (1975), and Williams and Lagakos (1977). A possible approach is to identify prognostic factors related to the causes of death under study. Consider, for example, a study of deaths from lung cancer and from heart disease among a group of workers exposed to a suspected health hazard. The two diseases may not be related among nonsmokers, or among smokers, even if they are related in the study group as a whole. The independent risks model could then be applied after stratification in the study group by smoking habit.

It must be stressed, however, that without additional data there is no way of statistically testing the validity of the independent risks model.

11.9 ALTERNATIVE TECHNIQUES

Although it is both simple and flexible, the proportional hazards model is not the only approach to the analysis of survival data. Three other methods are worthy of mention.

11.9.1 Modeling the Predicted Survival Time

Zippin and Armitage (1966) model the predicted survival time in the absence of censoring as a linear function of explanatory (risk and confounding) variables. If constant hazards are assumed, this amounts to expressing the reciprocal of the hazard function for each individual as a linear function of explanatory variables. A drawback of this approach is that it can lead to negative predicted survival times.

11.9.2 Additive Hazards

Further developing the idea of Section 11.8, we may think of the risk and confounding factors as acting additively on the hazard function rather than multiplicatively. Each explanatory variable introduces a "competing risk" which operates independently of the other competing risks, leading to a total hazard function of the form

$$\lambda(t) = \lambda^*(t) + \nu,$$

where ν depends on the explanatory variables and so varies from individual to individual, and $\lambda^*(t)$ is a hazard function common to all individuals.

Unfortunately, this model, although intuitively appealing, is hard to analyze, since the actual risk that results in death cannot be observed. Moreover, the possibility of negative estimated risks is still present.

Drolette (1966) discusses survival models with additive hazards.

11.9.3 Accelerated Life Testing

In accelerated life testing it is assumed that the explanatory variables act on the risk of death by speeding up or slowing down the time scale. It can be shown that this leads to a hazard function

$$\lambda(t) = \nu\lambda^*(\nu t) \tag{11.8}$$

with ν and $\lambda^*(t)$ defined as in Section 11.9.2. This differs from the proportional hazard model, which has

$$\lambda(t) = \nu\lambda^*(t) \tag{11.9}$$

It is easy to show however, that (11.8) and (11.9) define the same class of hazard functions if $\lambda^*(t)$ takes the special form

$$\lambda^*(t) = \alpha t^{\beta-1}$$

where α and β are unknown constants. This is known as the *Weibull hazard function* and is of considerable importance in industrial applications of survival analysis. These often involve testing components to destruction after subjecting them to a stress which is assumed to speed up the aging process—hence the name and technique of accelerated life testing.

Further discussion of acelerated life testing and Weibull hazard functions, including estimation of the constants α and β, is given by Mann et al. (1974), Gross and Clark (1975), and Kalbfleisch and Prentice (1979).

At least in applications where no theoretical form for survival distributions can be assumed, as is usual for studies involving human subjects, techniques based on the proportional hazards model appear superior to those introduced in this section. As shown by Kalbfleisch (1974), Efron (1977), and Oakes (1977), the relative efficiency of Cox's regression approach, compared with likelihood methods that assume a fully parametric form for the hazard function, is generally good.

APPENDIX 11A   SURVIVAL ANALYSIS IN CONTINUOUS
AND DISCRETE TIME

Using the notation of conditional probability, the discrete time hazard function defined in Section 11.2 may be written

$$\lambda(t) = \text{Pr } \{\text{death between } t \text{ and } t + 1 \mid \text{survival until } t\}.$$

The survivor function

$$S(t) = \text{Pr } \{\text{survival beyond } t\}$$

is related to $\lambda(t)$ by the equation

$$S(t + 1) = \prod_{i = 0}^{t} [1 - \lambda(i)].$$

The use of the discrete hazard in the discussion of the proportional hazard model (Section 11.3) is not quite correct theoretically. If the survivor function $S(t)$ is continuous, it is appropriate to define $\lambda(t)$, for all real t, as a hazard *rate* or *density*,

$$\lambda(t) = \lim_{h \to 0} \frac{1}{h} \text{Pr } \{\text{death between } t \text{ and } t + h \mid \text{survival until } t\}.$$

It can be shown that, with $S(t)$ defined as above,

$$S(t) = \exp \left[- \int_0^t \lambda(u) \, du \right].$$

In continuous time, the proportional hazard model of Section 11.3 is still

$$\lambda_1(t) = \theta \lambda_0(t),$$

that is, as given in Section 11.3 but with the different interpretations of $\lambda_0(t)$ and $\lambda_1(t)$. The relation

$$S_1(t) = [S_0(t)]^\theta$$

between the survivor functions of the two groups is now exact.

For data in discrete time, the original definition of $\lambda(t)$ in Section 11.2 is retained, but the relation between the hazard functions in the two groups is modified to

$$\frac{\lambda_1(t)}{1 - \lambda_1(t)} = \frac{\theta \lambda_0(t)}{1 - \lambda_0(t)}$$

to aid in fitting the model (Cox, 1972). In this model θ is interpreted as the odds ratio for the treatment effect rather than the relative risk. However, the treatment effect θ as defined here will be close to that introduced in Section 11.3 provided that the grouping intervals are not too wide.

APPENDIX 11B THE KAPLAN–MEIER (PRODUCT-LIMIT) ESTIMATOR

Kaplan and Meier (1958) give a simple estimator of the underlying survivor function $S(t)$ for the survival time of a homogeneous group of subjects, from censored data, which, unlike the actuarial (life-table) estimator introduced in Section 11.2, does not require preliminary calculation of the $\hat{\lambda}(t)$. Let the observed times of death be denoted by $t_1 < t_2 < \cdots < t_n$, and suppose that there are exactly r_i individuals who are known to survive until time t_i, of whom m_i die at time t_i. The estimated survivor function is defined as

$$\hat{S}(t) = \prod_{l=1}^{i} \left(1 - \frac{m_l}{r_l}\right) \qquad t_i \le t < t_i + 1.$$

The Kaplan–Meier estimator, also known as the *product-limit estimator*, will usually be numerically similar to the actuarial estimator, but because of the arbitrary nature of the grouping needed to calculate the actuarial estimator, it possesses certain advantages. Breslow and Crowley (1974) point out that the actuarial estimator does not become unbiased in large samples, although they suggest that the bias is not likely to be serious unless deaths occur in fewer than 10 of the grouping intervals used to calculate the $\hat{\lambda}(t)$.

The Kaplan–Meier estimator is approximately unbiased in large samples provided that the survival times t_i are reported exactly (i.e., without rounding or truncation).

APPENDIX 11C COX'S REGRESSION MODEL

The regression model introduced by Cox (1972) specifies the hazard rate $\lambda_i(t, \mathbf{X}_i)$ for the ith subject in terms of a vector of covariates $\mathbf{X}_i(t) = (X_{i1}(t), X_{i2}(t), \ldots, X_{ip}(t))$ specific to that individual and a vector $\beta = (\beta_1, \beta_2, \ldots, \beta_p)$ of regression parameters common to all subjects, as

$$\lambda_i(t, \mathbf{X}_i) = \lambda^*(t) \exp\left[\sum_{r=1}^{p} \beta_r X_{ir}(t)\right] = \lambda^*(t) e^{\beta \mathbf{X}_i(t)}$$

where $\lambda^*(t)$ is an unknown function of time. This generalizes the proportional hazards model by replacing the relative risk θ by the exponential of a linear predictor which may depend on many different covariates. Model (11.7) of Section 11.4 is recovered by taking $p = 2$ and the two components of $\mathbf{X}$ to be the treatment indicator variable (R) and the confounding variable.

The ratio of the hazard functions of two subjects, say the first and the second, becomes

$$\exp\left\{\sum_r \beta_r [X_{1r}(t) - X_{2r}(t)]\right\}. \tag{11.10}$$

If, as often happens, the $X_{ir}(t)$ do not depend on time but are constant for each subject, the hazard ratio (11.10) is also not dependent on time.

Cox (1972, 1975) has shown how estimates of the coefficients $\beta_1, \beta_2, \ldots, \beta_p$ may be obtained. Let t_i denote the time of death of patient i, let the patients be ordered so that $t_1 < t_2 < \cdots < t_n$, and suppose that there are no ties (otherwise, a slightly more complicated procedure is needed). Let R_i denote the set of individuals whose survival or censoring times equal or exceed t_i. Then the conditional probability that subject i dies at time t_i, given that exactly one subject from R_i dies at time t_i, is

$$\frac{\lambda_i(t_i, \mathbf{X}_i)}{\sum_{j \in R_i} \lambda_j(t_i, \mathbf{X}_j)} = \frac{e^{\beta \mathbf{X}_i(t)}}{\sum_{j \in R_i} e^{\beta \mathbf{X}_j(t)}}. \tag{11.11}$$

The estimate $\hat{\beta}$ of the vector β is chosen to maximize

$$L = \sum_{i=1}^{n} \beta \mathbf{X}_i(t) - \sum_{i=1}^{n} \ln \left[\sum_{j \in R_i} e^{\beta \mathbf{X}_j(t)} \right],$$

the sum of the logarithms of (11.11). Efficient computer programs can achieve this quite rapidly provided that p and n are not too large.

Cox (1975) has indicated that L has the large-sample properties of a log-likelihood function, so that the estimates $\hat{\beta}$ of β are asymptotically normally distributed with mean β and covariance matrix given by the inverse of the expectation of the matrix of second derivatives of L with respect to β. Efron (1977) and Oakes (1977) have shown how this expectation can in principle be calculated. However, it is much simpler to estimate it by the actual matrix of second derivatives of L, calculated at the observed maximum $\beta = \hat{\beta}$. The square roots of the diagonal elements of the inverse of this matrix give standard errors for the estimated parameters β_r. To test the null hypothesis $H_0: \beta_1 = \beta_2 = \cdots = \beta_s = 0$, we can calculate the ratio ν of the unconstrained maximum likelihood to the likelihood maximized under the constraint $\hat{\beta}_1 = \hat{\beta}_2 = \cdots = \hat{\beta}_s = 0$ and refer $2 \ln \nu$ to the chi-square distribution with s degrees of freedom.

APPENDIX 11D   BRESLOW'S ESTIMATOR OF THE SURVIVOR FUNCTION IN COX'S MODEL

In this appendix we shall assume that no covariate function depends on time. Under Cox's model the distribution function of the survival time of a subject with $X_{i1} = X_{i2} = \cdots = X_{ip} = 0$ is

$$F^*(t) = 1 - \exp \left[- \int_0^t \lambda^*(u) \, du \right] = 1 - S^*(t).$$

For a subject with a nonzero value $\mathbf{X}$ of the vector of covariates, the distribution function is $F(t) = 1 - S(t)$, where

$$S(t) = \exp\left[- \int_0^t \lambda(u, X)du\right]$$

$$= \exp\left[- e^{\beta X} \int_0^t \lambda^*(u) \, du\right]$$

$$= [S^*(t)]^\alpha,$$

where $\alpha = e^{\beta X}$.

As discussed in Section 11.5, we may wish to estimate these survivor functions after having obtained an estimate $\hat{\beta}$ of β. Breslow (1974) suggests that

$$\hat{S}^*(t_i) = \prod_{l=1}^{i} \left[1 - \frac{m_l}{\sum_{j \in R_l} \exp(\hat{\beta}X_j)}\right],$$

where, as in Appendix 11C, $t_1 < t_2 < \cdots < t_n$ denote the ordered times of death, and R_i and m_i denote the risk set and number of deaths at time t_i, respectively. Breslow suggests interpolation between successive t_i, making $\hat{S}^*(t)$ a continuous function of t, but it seems more natural to preserve the close analogy with the Kaplan–Meier estimate (Appendix 11B) for a homogeneous group of subjects and write

$$\hat{S}^*(t) = \hat{S}^*(t_i) \qquad t_i \leq t < t_{i+1}.$$

For a nonzero value of X, the corresponding estimate is

$$\hat{S}(t) = [\hat{S}^*(t)]^{\hat{\alpha}},$$

with $\hat{\alpha} = e^{\hat{\beta}X}$.

The sampling theory of Breslow's estimator has recently been investigated by Tsiatis (1978a, b).

REFERENCES

Aalen, O. (1976), Nonparametric Inference with Multiple Decrement Models, *Scandinavian Journal of Statistics*, **3**, 15–27.

Armitage, P. (1971), *Statistical Methods in Medical Research*, London: Blackwell.

Bradford Hill, A. (1977), *A Short Textbook of Medical Statistics*, London: Hodder and Stoughton.

Breslow, N. (1970), A Generalized Kruskal-Wallis Test for Comparing K Samples Subject to Unequal Patterns of Censorship, *Biometrika*, **57**, 579–594.

Breslow, N. (1972), Contribution to the discussion of Cox (1972), cited below.

Breslow, N. (1974), Covariance Analysis of Censored Survival Data, *Biometrics*, **30**, 89–99.

Breslow, N. (1975), Analysis of Survival Data under the Proportional Hazards Model, *International Statistical Review*, **43**, 45–58.

Breslow, N., and Crowley, J. (1974), A Large Sample Study of the Life-Table and Product Limit Estimates under Random Censorship, *Annals of Statistics*, **2**, 437–453.

Chiang, C. L. (1961a), A Stochastic Study of the Life-Table and Its Applications III. The Follow-Up Study with the Consideration of Competing Risks, *Biometrics*, **77**, 57–78.

Chiang, C. L. (1961b), On the Probability of Death from Specific Causes in the Presence of Competing Risks, *Proceedings of the Fourth Berkeley Symposium on Mathematical Statistics and Probability*, Vol. 4, Berkeley, CA: University of California Press, pp. 169–180.

Chiang, C. L. (1970), Competing Risks and Conditional Probabilities, *Biometrics*, **26**, 767–776.

Colton, T. (1974), *Statistics in Medicine*, Boston: Little, Brown.

Cox, D. R. (1953), Some Simple Tests for Poisson Variates, *Biometrika*, **40**, 354–360.

Cox, D. R. (1959), The Analysis of Exponentially Distributed Lifetimes with Two Types of Failure, *Journal of the Royal Statistical Society, Series B*, **21**, 411–421.

Cox, D. R. (1972), Regression Models and Life-Tables, *Journal of the Royal Statistical Society, Series B*, **34**, 187–202.

Cox, D. R. (1975), Partial Likelihoods, *Biometrika*, **62**, 599–607.

Cox, D. R. (1978), Some Remarks on the Role in Statistics of Graphical Methods, *Applied Statistics*, **27**, 4–9.

Cox, D. R. (1979), A Note on the Graphical Analysis of Survival Data, *Biometrika*, **66**, 188–190.

Cox, D. R., and Snell, E. J. (1968), A General Definition of Residuals, *Journal of the Royal Statistical Society, Series B*, **30**, 248–275.

Crowley, J., and Hu, M. (1977), The Covariance Analysis of Heart Transplant Data, *Journal of the American Statistical Association*, **72**, 27–36.

Drolette, M. E. (1966), *Exponential Models for Survival in Chronic Disease*, Ph.D. thesis, Harvard University.

Efron, B. (1967), The Two-Sample Problem with Censored Data, *Proceedings of the Fifth Berkeley Symposium on Mathematical Statistics and Probability*, Vol. 4, Berkeley, CA: University of California Press, pp. 831–854.

Efron, B. (1977), The Efficiency of Cox's Likelihood Function for Censored Data, *Journal of the American Statistical Association*, **72**, 557–565.

Fisher, L., and Kanarek, P. (1974), Presenting Censored Survival Data When Censoring and Survival Times May Not Be Independent, *in* E. Proschan and R. J. Serfling, Eds., *Reliability and Biometry*, Society of Industrial and Applied Mathematics, 1974, pp 303–326.

Freireich, E. J., Gehan, E., Frei, E., Schroeder, L. R., Wolman, R. A., Burgert, O. E., Mills, S. D., Pinkel, D., Selawry, O. S., Moon, J. H., Gendel, B. R., Spurr, C. L., Storrs, R., Haurani, F., Hoogstraten, B., and Lee, S. (1963). The Effect of 6-Mercaptopurine on the Duration of Steroid-induced Remissions in Acute Leukemia: A Model for Evaluation of Other Potentially Useful Therapy, *Blood*, **21**, 699–716.

Gail, M. H. (1972), Does Cardiac Transplantation Prolong Life? A Reassessment, *Annals of Internal Medicine*, **76**, 815–817.

Gail, M. H. (1975), A Review and Critique of Some Models in Competing Risk Analysis, *Biometrics*, **31**, 209–222.

Gehan, E. A. (1965a), A Generalized Wilcoxon Test for Comparing Arbitrarily Single-Censored Samples, *Biometrika*, **52**, 203–223.

Gehan, E. A. (1965b), A Generalized Two-Sample Wilcoxon Test for Doubly Censored Data, *Biometrika*, **52**, 650–652.

Gillespie, M. J. and Fisher, L. (1979) Confidence Bands for the Kaplan-Meier Survival Curve Estimate, *Annals of Statistics*, **7**, 920-924.

Greenwood, M. (1926), The Errors of Sampling of the Survivorship Tables, Appendix 1, *Reports on Public Health and Statistical Subjects 33*, London: Her Majesty's Stationery Office.

Gross, A. J., and Clark, V. A. (1975), *Survival Distributions, Reliability Applications in the Biomedical Sciences*, New York: Wiley.

Kalbfleisch, J. D. (1974), Some Efficiency Calculations for Survival Distributions, *Biometrika*, **61**, 31-38.

Kalbfleisch, J. D., and Prentice, R. L. (1973), Marginal Likelihoods Based on Cox's Regression and Life Model, *Biometrika*, **60**, 267-278.

Kalbfleisch, J. D. and Prentice, R. L. (1979) *The Statistical Analysis of Failure Time Data*, New York: Wiley. In press.

Kaplan, E. L., and Meier, P. (1958), Nonparametric Estimation from Incomplete Observations, *Journal of the American Statistical Association*, **53**, 457-481.

Kay, R. (1977), Proportional Hazard Regression Models and the Analysis of Censored Survival Data, *Applied Statistics*, **26**, 227-237.

Lagakos, S. W. (1978), A Covariate Model for Partially Censored Data Subject to Competing Causes of Failure, *Applied Statistics*, **27**, 235-241.

Lancaster, T. (1978), Econometric Methods for the Duration of Unemployment, Hull economic research paper, University of Hull, England.

Liddell, F. D. K., McDonald, J. C., and Thomas, D. C. (1977), Methods of Cohort Analysis Appraisal by Application to Asbestos Mining, *Journal of the Royal Statistical Society, Series A*, **140**, 469-491.

Mann, N. R., Schafer, R. E., and Singpurwalla, N. D. (1974), *Methods for Statistical Analysis of Reliability and Life Data*, New York: Wiley.

Mantel, N. (1966), Evaluation of Survival Data and Two New Rank Order Statistics Arising in Its Consideration, *Cancer Chemotherapy Reports*, **50**, 163-170.

Mantel, N. (1967), Ranking Procedures for Arbitrarily Restricted Observations, *Biometrics*, **23**, 65-78.

Mantel, N., and Byar, D. P. (1974), Evaluation of Response Time Data Involving Transient States; An Illustration Using Heart Transplant Data, *Journal of the American Statistical Association*, **69**, 81-86.

Meier, P. (1975), Estimation of a Distribution Function from Incomplete Observations, *in* J. Gani, Ed., *Perspectives in Probability and Statistics*, London: Academic Press, pp 67-87.

Nelson, W. (1972), Theory and Application of Hazard Plotting for Censored Failure Data, *Technometrics*, **14**, 945-965.

Oakes, D. (1972), Contribution to the discussion of Cox (1972), cited above.

Oakes, D. (1977), The Asymptotic Information in Censored Survival Data *Biometrika*, **64**, 441-448.

Peto, R. (1972), Rank Tests of Maximal Power against Lehmann-Type Alternatives, *Biometrika*, **59**, 472-474.

Peto, R., and Peto, J. (1972), Asymptotically Efficient Rank Invariant Test Procedures, *Journal of the Royal Statistical Society, Series A*, **135**, 185-207.

Peto, R., and Pike, M. C. (1973), Conservatism of the Approximation $\Sigma(O - E)^2/E$ in the Logrank Test for Survival Data or Tumour Incidence Data, *Biometrics*, **29**, 579-584.

Peto, R., Pike, M. C. Armitage, P. Breslow, N. E. Cox, D. R. Howard, S. V. Mantel, N. McPherson,

K., Peto, J., and Smith, P. G. (1976), Design and Analysis of Randomized Clinical Trials Requiring Prolonged Observation of Each Patient, I. Introduction and Design, *British Journal of Cancer,* **34,** 585–612.

Peto, R., Pike, M. C., Armitage, P., Breslow, N. E., Cox, D. R., Howard, S. V., Mantel, N., McPherson, K., Peto, J., and Smith, P. G. (1977), Design and Analysis of Randomized Clinical Trails Requiring Prolonged Observation of Each Patient, II. Analysis and Examples, *British Journal of Cancer,* **35,** 1–39.

Snedecor, G. W., and Cochran, W. G. (1967), *Statistical Methods,* 6th ed., Ames, IA: Iowa State University Press.

Tarone, R. E., and Ware, J. (1977), On Distribution Free Tests for the Equality of Survival Distributions, *Biometrika,* **64,** 156–160.

Tsiatis, A. (1975), A Nonidentifiability Aspect of the Problem of Competing Risks, *Proceedings of the National Academy of Sciences,* **27,** 20–22.

Tsiatis, A. (1978a), A Heuristic Estimate of the Asymptotic Variance of the Survival Probability in Cox's Regression Model, University of Wisconsin Technical Report.

Tsiatis, A. (1978b), A Large Sample Study of the Estimate for the Integrated Hazard Function in Cox's Regression Model for Survival Data, University of Wisconsin Technical Report.

Turnbull, B. W., Brown, B. W., and Hu, M. (1974), Survivorship Analysis of Heart Transplant Data, *Journal of the American Statistical Association,* **69,** 74–80.

Williams, J. S., and Lagakos, S. W. (1977), Models for Censored Survival Analysis: Constant and Variable-Sum Models, *Biometrika,* **64,** 215–224.

Zippin, C., and Armitage, P. (1966), Use of Concomitant Variables and Incomplete Survival Information in the Estimation of an Exponential Survival Parameter, *Biometrics,* **22,** 665–672.

CHAPTER 12

Analyzing Data from
Premeasure/Postmeasure Designs

The typical premeasure/postmeasure design consists of an assessment for each individual on the outcome variable prior to the treatment (the premeasure) and a remeasure of each case on the same outcome variable after the treatment (the postmeasure). For example, in medical research we might wish to compare a new drug therapy for hypertension with the conventional treatment. It seems natural to measure the blood pressure for each subject in both treatment and comparison groups prior to the therapy and again some time later (after, say, 3 months of treatment). If subjects are not randomly assigned to groups, ap-

235

parent outcome differences between the groups in blood pressure level may be attributable to the alternative drug programs, to differences in blood pressure prior to the treatment, or to other factors confounded with group membership.

In one sense, the availability of a premeasure presents no new conceptual issues. The premeasure is simply a variable to be considered for use in the adjustment procedures introduced in Chapters 6 to 10. On the other hand, the fact that the premeasure represents a pretreatment observation on the outcome variable suggests that it ought to be accorded some special status. An observed difference across groups on the premeasure constitutes evidence of relevant group differences prior to treatment. In order to assess the effect of the treatment, the analysis must adjust for this confounding factor. Moreover, because of the premeasure's unique relationship to the outcome, it would seem to be an ideal variable for use in statistical adjustment.

The Head Start Planned Variation Study (HSPV) (see Smith, 1973; Weisberg, 1973) provides a good example of a premeasure/postmeasure design that we will employ throughout this chapter. The purpose of the HSPV study was to examine the impact of alternative preschool curricular models on child development. Table 12.1 presents some data from the study on one of the main outcome variables, the Pre-School Inventory.* There is a mean difference of 4.4 points between the "innovative" curriculum model and the comparison group (a sample of children from ordinary Head Start programs) at the postmeasure. This reflects a possible program effect. Notice, however, that there is a premeasure mean difference of 2.5 points. This suggests a possible important difference between the groups prior to the intervention. In such a case, it is not clear what part of the 4.4-point difference, if any, is attributable to the program.

In order to generate a valid standard of comparison for assessing the effects of the innovation, we must somehow adjust for premeasure differences across groups. How to utilize most effectively the premeasure data to reduce the bias in estimating treatment effects is the major theme of this chapter. Our discussion here will be somewhat different from that of previous chapters. Rather than emphasizing new methodological approaches, this chapter focuses more on developing a basic understanding of the special nature of the confounding problem in premeasure/postmeasure designs, and on identifying those situations where existing methods, such as those already introduced in Chapters 6 to 10, can be usefully applied. We conclude this chapter with a brief discussion of some recent developments based on individual subject growth curve models.

---

* The Pre-School Inventory measures the general knowledge of pre-school-age children. It focuses on a diverse set of information that most children should possess by school entry (e.g., "What is your name?"). It consists of 32 items and is scored in terms of number of items correct. Thus scores can range from 0 to 32.

Table 12.1   Data from Head Start Planned Variation Study

	Premeasure	Postmeasure	Correlation between Premeasure and Postmeasure
Innovative curriculum model	$\overline{Y}_1(t_1) = 17.1$ s.d. = 6.1 $n = 157$	$\overline{Y}_1(t_2) = 23.3$ s.d. = 4.6	$r = .67$
Standard Head Start programs	$\overline{Y}_0(t_1) = 14.6$ s.d. = 6.2 $n = 669$	$\overline{Y}_0(t_2) = 18.9$ s.d. = 5.8	$r = .78$
	$\overline{Y}_1(t_1) - \overline{Y}_0(t_1) = 2.5$	$\overline{Y}_1(t_2) - \overline{Y}_0(t_2) = 4.4$	

12.1 REVIEW OF NOTATION

We need to introduce at this point a notation reflecting the temporal nature of premeasure/postmeasure data. Let us define

$\overline{Y}_1(t_1)[\overline{Y}_0(t_1)]$ as the observed mean for the treatment (control) group on the premeasure taken at time t_1

$\overline{Y}_1(t_2)[\overline{Y}_0(t_2)]$ as the observed mean for the treatment (control) group on the postmeasure taken at time t_2

$\overline{Y}_1^*(t_2)$ as the postmeasure mean that would have been observed in the program group had the treatment not been applied.

Although, in principle, we cannot measure $\overline{Y}_1^*(t_2)$, this entity is a useful heuristic for thinking about the standard of comparison problem in premeasure/postmeasure designs. We can now define the treatment effect, α, as the expected difference between the observed postmeasure mean for the treatment group and the postmeasure mean for this group that would have been observed in the absence of the treatment:

$$\alpha = E[\overline{Y}_1(t_2) - \overline{Y}_1^*(t_2)]. \qquad (12.1)$$

In premeasure/postmeasure designs, $E[\overline{Y}_1^*(t_2)]$ represents the ideal standard of comparison for assessing the effects of the treatment. Note, if subjects had been randomly assigned to groups, we would have

$$E[\overline{Y}_0(t_2)] = E[\overline{Y}_1^*(t_2)]. \qquad (12.2)$$

As a result, in randomized studies the simple unadjusted postmeasure mean difference, $\overline{Y}_1(t_2) - \overline{Y}_0(t_2)$, is an unbiased estimate of the treatment effect, α.

Because of confounding on the premeasure, however, (12.2) does not hold. We examine in the next section several alternative strategies that have been frequently used to approximate the ideal represented in (12.1) and (12.2).

To simplify the notation in this chapter, we assume large samples. As a result, we can drop the expectation symbols and refer simply to the sample statistics such as $\overline{Y}_1(t)$ and $\overline{Y}_0(t)$. The use of a more precise notation would only complicate the presentation, but would not alter the results.

12.2   TRADITIONAL APPROACHES TO ESTIMATING TREATMENT EFFECTS IN PREMEASURE/ POSTMEASURE DESIGNS

The simple gain or change score is the most natural approach to analyze the pre- and postmeasure data. We compute the mean gain for individuals in the treatment and control groups, and define the estimated treatment effect, $\hat{\alpha}$, as the difference in these mean gains:

$$\hat{\alpha} = [\overline{Y}_1(t_2) - \overline{Y}_1(t_1)] - [\overline{Y}_0(t_2) - \overline{Y}_0(t_1)]. \tag{12.3}$$

For the HSPV data from Table 12.1,

$$\hat{\alpha}_{\text{simple gain}} = (23.3 - 17.1) - (18.9 - 14.6) = 1.9 \text{ points.}$$

Our estimate of the effect of the innovative curriculum is still positive, but considerably smaller than was suggested by simple inspection of the postmeasure means.

The gain score has a long history in social science research, much of it tied up with problems of "measurement of change" (see Harris, 1967). Considerable interest has focused on estimating individual change over time and relating these changes to characteristics of individuals, such as their premeasure score and other background variables. Because of errors of measurement in the change scores, the estimation problem can be quite difficult, and interest in this problem persists (see Blomquist, 1977). Some other techniques for the analysis of premeasure/postmeasure data—repeated measure ANOVA, and standardized gain scores—trace their existence back to this measurement of change literature. We note that with only two data points, the repeated measures ANOVA (see Winer, 1971) is mathematically equivalent to the simple gain score (12.3).

Each of these techniques for analyzing data from premeasure/postmeasure designs falls under the general heading of *linear adjustments*. These methods share the following form:

$$\hat{\alpha} = \overline{Y}_1(t_2) - \overline{Y}_0(t_2) - \hat{\beta}[\overline{Y}_1(t_1) - \overline{Y}_0(t_1)], \tag{12.4}$$

where $\hat{\alpha}$ is the treatment effect estimate. In each case, we adjust the postmeasure mean difference between groups, $[\overline{Y}_1(t_2) - \overline{Y}_0(t_2)]$, by some multiplier, $\hat{\beta}$, of the premeasure mean difference, $[\overline{Y}_1(t_1) - \overline{Y}_0(t_1)]$, between groups.

The adjustment coefficient $\hat{\beta}$ characterizes the particular linear adjustment method. Simple gain scores, for example, set $\hat{\beta}$ equal to 1. This can easily be seen by substituting a value of 1 for $\hat{\beta}$ in (12.4) and reorganizing the terms to get (12.3).

Standardized gain scores are somewhat more empirical in that they use characteristics of the premeasure and postmeasure data to determine $\hat{\beta}$. For standardized gain scores,

$$\hat{\alpha} = \overline{Y}_1(t_2) - \overline{Y}_0(t_2) - \frac{s_2}{s_1}[\overline{Y}_1(t_1) - \overline{Y}_0(t_1)], \qquad (12.5)$$

where

s_1 = pooled within-group estimated standard deviation of premeasure
s_2 = pooled within-group estimated standard deviation of postmeasure.

The adjustment coefficient for standardized gain scores is simply the ratio of the post- and premeasure standard deviations. If the variability within groups on the outcome measure changes over the time between pre- and postmeasurement, this is reflected in the adjustment coefficient. For some specific situations described in Section 12.5, this property makes standardized gains a useful strategy. Applying (12.5) to the HSPV data, we compute

$$\hat{\alpha}_{\text{std gains}} = (23.3 - 18.9) - \frac{5.6}{6.2}(17.1 - 14.6) = 2.1 \text{ points.}$$

The differences in results across the two types of gains analyses simply reflect the different values for $\hat{\beta}$ in each adjustment strategy.

More recent efforts at analyzing data from premeasure/postmeasure designs (see Linn and Slinde, 1977; Reichardt, 1979) argue that the premeasure should not be granted special status, as in gain score methods, but rather simply treated as another confounding variable. In principle, any technique already described in the earlier chapters could be appropriate, depending of course on the particular types of variables present in a given application. For the HSPV example, where both the pre- and postmeasure are numerical variables, analysis of covariance would be the method of choice. Using the basic analysis of covariance (AN-COVA) model from Chapter 8, we have

$$Y(t_2) = \mu + \alpha + \beta Y(t_1) + e. \qquad (12.6)$$

As demonstrated in Chapter 8, this leads to an estimate of the treatment effect, α, as

$$\hat{\alpha} = \overline{Y}_1(t_2) - \overline{Y}_0(t_2) - \hat{\beta}[\overline{Y}_1(t_1) - \overline{Y}_0(t_1)], \qquad (12.7)$$

where $\hat{\beta}$ is the pooled within-group regression coefficient of the postmeasure on the premeasure (Appendix 8A). Note that (12.7) is another example of a linear adjustment method.

Applying ANCOVA to the HSPV data, we compute

$$\hat{\alpha}_{\text{ANCOVA}} = (23.3 - 18.9) - (.76)(17.1 - 14.6) = 2.5 \text{ points.}$$

Thus, we have a third estimate of the treatment effect. While all are positive, they cover a range from 1.9 points (simple gain scores) to 2.5 points (ANCOVA). Each technique resulted in a different amount of adjustment for the initial premeasure differences between groups. Each used the observables, $\overline{Y}_0(t_2)$, $\overline{Y}_1(t_1)$, $\overline{Y}_0(t_1)$, in slightly different ways, to approximate the ideal but unobservable $\overline{Y}_1^*(t_2)$, the posttest mean for the treatment group that would have occurred in the absence of the intervention.

12.3 THE BASIC PROBLEM

In trying to evaluate the techniques mentioned above, we must attend to a distinctive feature of premeasure/postmeasure designs. Subjects are often growing or changing on the outcome measure even in the absence of a treatment. Further, individual subjects may be growing or changing at different rates. Depending upon the nature of the individual change phenomena, and how individuals are assembled into treatment and comparison groups, this can profoundly affect the bias-reducing properties of each strategy. We illustrate with two examples in this section the problems caused when subjects are growing at different rates. A more formal treatment is offered in Section 12.4.

Campbell and Erlebacher (1970), in their critique of the Westinghouse-Ohio evaluation of Head Start, suggest that in this particular premeasure/postmeasure data set the mean difference between groups on the premeasure is probably the result of differential mean growth rates prior to the premeasure point. They hypothesize that this differential mean growth between groups simply reflects the different growth rates of individuals within the groups. The mean gap between the groups should increase over time as the individuals within the groups spread farther apart. Assuming that, in the absence of a treatment, the differential individual growth rates would continue undisturbed results in the "fan spread" hypothesis illustrated in Figure 12.1. The asterisks represent the data observed by the evaluators. The dashed lines represent the hypothesized mean growth trends for each group in the absence of an intervention. Thus we have the data given in Table 12.2.

The true treatment effect is reflected in

$$\overline{Y}_1(t_2) - \overline{Y}_1^*(t_2) = 27.0 - 23.0 = 4.0 \text{ points.}$$

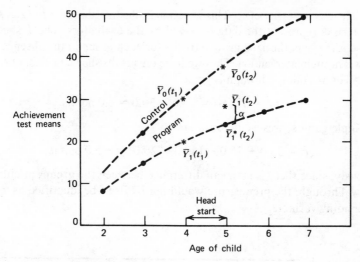

Figure 12.1 Fan spread hypothesis.

Table 12.2 Data from Fan Spread Illustration

$\overline{Y}_1(t_1) = 19.0$	$\overline{Y}_1(t_2) = 27.0$	$\overline{Y}_1^*(t_2) = 23.0$
$\overline{Y}_0(t_1) = 30.0$	$\overline{Y}_0(t_2) = 38.0$	

If we computed a simple gain score, however, the estimated treatment effect would be

$$\hat{\alpha} = 27.0 - 19.0 - (38.0 - 30.0) = 0 \text{ points,}$$

and this analysis would result in a 4-point bias in favor of the comparison group, suggesting no treatment effect when in fact a positive effect had occurred.

A second example derives from the national evaluation of Project Follow Through*—a nonrandomized study of alternative curricula for the early elementary grades. Following a positive preschool (Head Start) experience, it was argued that well-designed elementary school programs were needed in order to solidify these early gains. The Follow Through curricula were developed in response. Many of the children in the Follow Through program group had had prior instruction in Head Start. Most of the control group had not received Head Start, however. The premeasure data were collected at entry into Follow Through, and so include for the Follow Through children the program effects

---

* The example as presented here somewhat oversimplifies the data analysis difficulties in the Follow Through evaluation. We present only a small segment that is useful for this illustration.

of the Head Start experience. This situation is illustrated in Figure 12.2. Again, the asterisks represent the data observed by the evaluators. The dashed lines represent the hypothesized growth trends for each group in the absence of the Follow Through intervention. Thus we have the results shown in Table 12.3. The true effect of Follow Through is

$$\overline{Y}_1(t_2) - \overline{Y}_1^*(t_2) = 35.0 - 30.0 = 5.0 \text{ points,}$$

but simple gains scores would estimate

$$\hat{\alpha} = 35.0 - 25.0 - (40.0 - 25.0) = -5.0 \text{ points.}$$

Curiously, since there is no mean difference between the groups at entry into Follow Through, the premeasure would not by itself be identified as a major confounding variable.

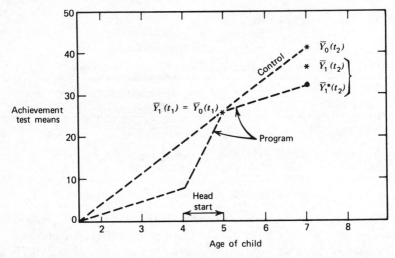

Figure 12.2 Problems associated with prior treatment effects—Head Start/Follow Through evaluation.

Table 12.3   Hypothetical Data based on Head Start/Follow Through Evaluation[a]

$\overline{Y}_1(t_1) = 25.0$	$\overline{Y}_1(t_2) = 35.0$	$\overline{Y}_1^*(t_2) = 30.0$
$\overline{Y}_0(t_1) = 25.0$	$\overline{Y}_0(t_2) = 40.0$	

[a] t_1 = premeasure time point at age 5; t_2 = postmeasure time point at age 7.

12.4 HEURISTIC MODEL FOR ASSESSING BIAS REDUCTION

The subjects in premeasure/postmeasure designs are often changing over time on the outcome variable even in the absence of a specific treatment. The pre- and postmeasure values for any subject represent two snapshots of a subject's growth trajectory. We can model the change on the outcome variable over time with an individual growth curve. The trajectories for these curves will vary across subjects, reflecting individual differences in growth rate. When we select subjects for a treatment and a comparison group, we create a distribution of growth curves for each group. If the selection process is nonrandom, however, the characteristics of the growth curve distribution may vary across groups. Specifically, the non-

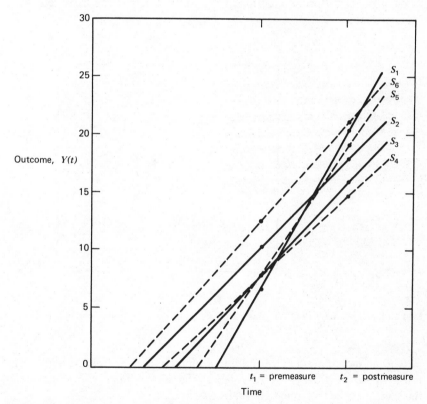

Figure 12.3a Individual growth curves in the context of a premeasure/postmeasure design (assuming no treatment effect). Solid lines, treatment group (S_1 to S_3); dashed lines, control group (S_4 to S_6).

random assignment of subjects to groups may result in differential mean growth trajectories across groups in the absence of any treatment. This constitutes the major confounding factor that analytic methods for premeasure/postmeasure designs must address. We can clarify this point through an illustration.

We present in Figure 12.3a a hypothetical set of individual growth curves. For simplicity we assume that each individual's growth over time can be adequately modeled by a straight line. The resulting pre- and postmeasure data are plotted in Figure 12.3b. If we assume that the first three subjects constitute the treatment group, and the remaining three the control, we can observe mean growth trajectories for the two groups in the absence of any treatment intervention. These are represented by the solid lines in Figure 12.4a. Notice that while on the average the treatment group starts out higher at the premeasure point, it has fallen behind the control group at the postmeasure point. Thus even in the absence of an intervention, it would appear that the treatment has a negative effect.

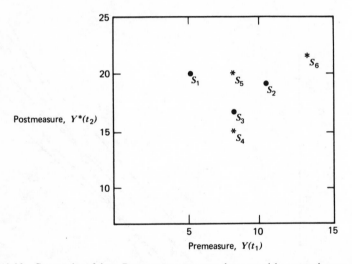

Figure 12.3b   Scatterplot of data. Dots, treatment group data; asterisks, control group data.

Now let us assume that we apply an intervention to the treatment group and it has a positive effect (as represented by the dashed line in Figure 12.4b): In this case the true treatment effect, α, is 3.5 points:

$$\alpha = \overline{Y}_1(t_2) - \overline{Y}_1^*(t_2) = 21.0 - 17.5 = 3.5 \text{ points.}$$

If we performed a simple gain score analysis, however, we find that

$$\hat{\alpha} = (21.0 - 18.5) - (9.5 - 7.0) = 0 \text{ points.}$$

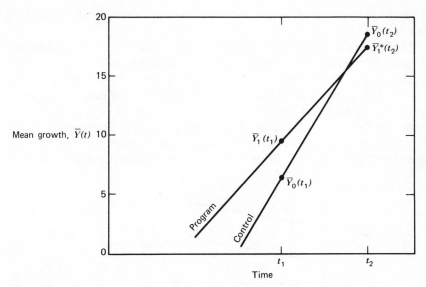

Figure 12.4a Mean growth trajectories in the absence of a treatment intervention.

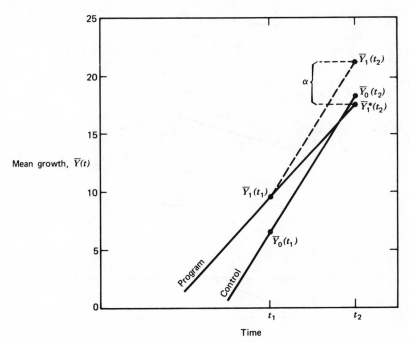

Figure 12.4b Mean growth trajectories with a positive treatment effect. Solid lines, mean growth trajectory in the absence of a treatment intervention; dashed line, mean growth trajectory for treatment group under the presence of a treatment effect α; α, true treatment effect.

Thus the gain score analysis underestimates the true treatment effects, resulting in a bias of 3.5 points.

Most generally, if we think simply of two curves, one each for a treatment and control group, representing the mean growth trajectories of the group over time in the absence of any intervention, one of four mean growth patterns must emerge: (a) the mean growth curves could be parallel over time; (b) the mean growth curves may close together as time passes (the fan close case); (c) the mean growth curves may be spreading apart as time passes (the fan spread case); and (d) in the duration between the pretest and posttest the mean growth curves may cross (the crossover case). These four situations are illustrated in Figure 12.5.

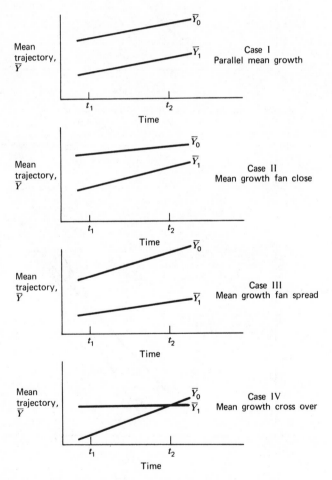

Figure 12.5   Four general classes of mean growth trajectories.

We should emphasize that these cases represent the expected growth trajectories for each group *in the absence of a treatment*.

Notice that while the initial mean difference between the groups, $\overline{Y}_1(t_1) - \overline{Y}_0(t_1)$, is the same in each case, the amount and direction of adjustment necessary varies across the four cases. If we limit consideration to linear adjustment methods [see (12.4)], it is easy to show that the theoretically correct adjustment coefficient, $\beta*$, is defined as

$$\beta* = \frac{\overline{Y}_1^*(t_2) - \overline{Y}_0(t_2)}{\overline{Y}_1(t_1) - \overline{Y}_0(t_1)}. \tag{12.8}$$

Note that $\beta*$ is simply the ratio of the expected postmeasure mean difference in the absence of the treatment to the observed premeasure mean difference.

Applying (12.8) to each of the four general cases defined in Figure 12.5, we see that

(1) For parallel mean growth, $\beta* = 1.0$.
(2) For mean growth fan close, $0 < \beta* < 1.0$.
(3) For mean growth fan spread, $\beta* > 1.0$.
(4) For mean growth crossover, $\beta* < 0$. (12.9)

Thus Figure 12.4 is an example of case 4. In the absence of an intervention, the mean growth curves would have crossed over. For our analysis to adjust appropriately when there is an intervention in the treatment groups, it requires a negative value for the coefficient β. With the use of simple gain scores (β set equal to 1), however, we actually adjusted in the wrong direction and created a more biased estimate of the treatment effect than if we had ignored the premeasure data! Only if the mean growth curves in the absence of a treatment had been parallel would the simple gain score approach have been correct.

In the context of a real premeasure/postmeasure study, however, we cannot directly apply (12.8) since we have no information on $\overline{Y}_1^*(t_2)$, the postmeasure mean for the treatment group that would have occurred in the absence of the intervention. Nevertheless, this equation, in combination with the four classes of mean growth trajectories in Figure 12.5, is helpful in understanding the appropriateness of a specific analysis strategy in a given case.

12.5 EXAMINING THE BEHAVIOR OF LINEAR ADJUSTMENTS

As we have already indicated, the simple gain score strategy is appropriate only under situations of parallel mean growth [case 1 of (12.9)]. The "stable state," where we expect no mean growth for either group in the absence of a treatment, represents a special instance from this class.

Consider again a study of therapies for hypertension. For most adults, blood pressure increases with age, so we have a potential "growth" problem. However, the change in blood pressure is gradual. If the hypertension study is such that the postmeasure, after drug therapy, is only 1 or 2 months later than the premeasure, on the average no change would be expected in the absence of the treatment. In this case, the use of a gain score analysis would be appropriate. Alternatively, if the study were of the effect of behavioral modification on blood pressure and there were, say, 5 to 10 years between the pre- and postmeasures, the growth trend problem would have to be considered.

As for linear adjustment methods other than gain scores, they all use, in one fashion or another, aspects of the across-subject variation and covariation in the pre- and postmeasures to estimate β. The variance in the pre- and postmeasure, and the covariance between them, are sensitive, however, to the nature of the individual growth curves and how they are distributed across groups. The importance of this can be seen in a rather extreme case.

Figure 12.6 contains two mean growth trajectories, each representing the

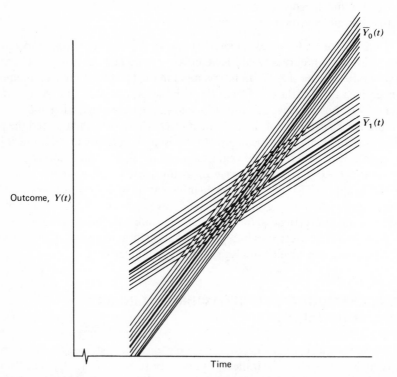

$\overline{Y}_0(t)$

$\overline{Y}_1(t)$

Outcome, $Y(t)$

Time

Figure 12.6 Differences between individual growth within groups and mean growth between groups. Single heavy line, mean growth; multiple light lines, individual growth.

aggregation of a set of parallel individual growth curves. Notice that in this very simple illustration the individuals maintain the same relative rank order, no matter where we slice the process. For any arbitrarily chosen t_1 and t_2 the pooled within-group correlation between the two measures is always 1.0, and the pooled within-group variance is constant. As a result, if we apply ANCOVA to such data, it can easily be shown that $\hat{\beta} = 1.0$ for any t_1 and t_2. A standardized gains analysis would also estimate the same value.

Notice, however, that the magnitude and direction of the bias depend on the specific choice of t_2. Further, the magnitude and direction of the appropriate adjustment coefficient depend upon the choice of both t_1 and t_2. We have reproduced the mean growth trajectories from Figure 12.6 in Figures 12.7a and 12.7b, where we consider different choices of t_1 and t_2. In Figure 12.7a the mean growth fan close implies that $\beta^* < 1.0$. The mean growth fan spread in Figure 12.7b requires that $\beta^* > 1.0$. Incidentally, in no case will the estimate, $\hat{\beta} = 1.0$, constitute the correct value!

In short, the bias that we are attempting to reduce is represented by the differences in the mean growth trajectories, which is essentially a *between-group*

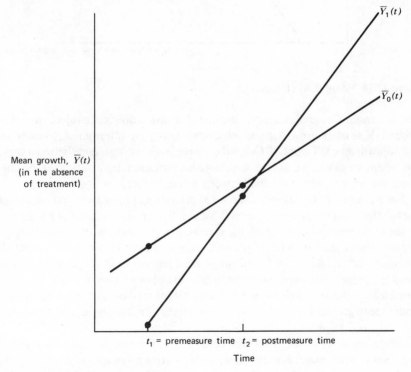

Figure 12.7a   Mean growth fan close.

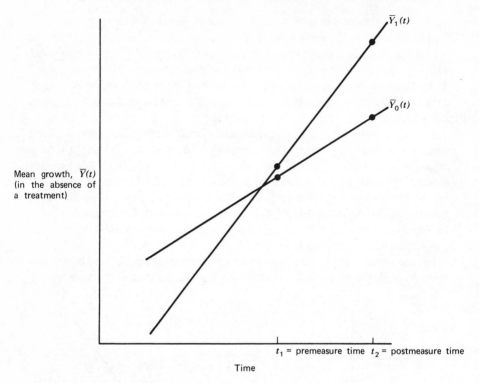

Mean growth, $\overline{Y}(t)$
(in the absence of
a treatment)

t_1 = premeasure time t_2 = postmeasure time

Time

Figure 12.7b   Mean growth fan spread.

phenomenon. The analysis methods, however, use individual subject variation around these mean growth trajectories, or *within-group* information, to estimate the adjustment coefficient, β. Only when there is some congruence between these two phenomena can we use the within-group variation/covariation on the outcome measure to adjust for the between-group differences.

For example, in typical educational applications, premeasure/postmeasure correlations usually range between 0.5 and 0.9. Further, because of the nature of the tests commonly employed, the variances in premeasures and postmeasures are often quite similar. As a result, we usually find that the β estimated through methods such as ANCOVA will range between values of 0.5 and 0.9. Thus, despite the fact that the mean growth fan spread or crossover might be a reasonable hypothesis in a given case, the adjustment methods are appropriate only under mean growth fan close. It can be shown that similar results also occur with standardized gains.

There are some situations where consistency of between- and within-group phenomena can occur. For example, under a strong fan spread model, where the mean growth trajectories (in the absence of the intervention) are spreading apart at the same rate as the trajectories within groups are spreading apart (see Figure 12.8), standardized gains should be an effective bias-reducing strategy (see Kenny, 1975). We could also postulate a corresponding strong fan close model, where the mean growth curves are closing together at the same time as the individual trajectories within each group are coming together. Standardized gains should be effective here, too.

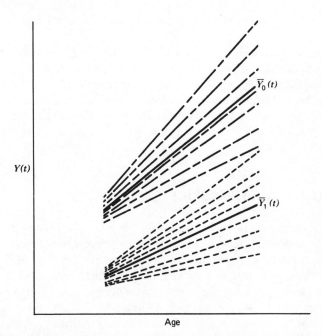

Figure 12.8 Strong fan spread model. Longer dash lines, individual growth curves for subjects in the control group; shorter dash lines, individual growth curves (in the absence of an intervention) for subjects in the treatment group.

More formally, Weisberg (1979) has shown that only under a very special case will ANCOVA totally eliminate the bias resulting from differential mean growth trajectories. For this to occur, the premeasure must be perfectly correlated with the slope of the individual trajectory that would have occurred between premeasure and postmeasure in the absence of treatment. In this case, the mean growth trajectory for each group is just a linear function of the premeasure mean.

Bryk and Weisberg (1977) have investigated the bias-reduction properties of several linear adjustment strategies under models of linear individual growth such as those illustrated previously in Figure 12.3. Similarly, Bryk (1977) has explored some nonlinear individual growth curve models. It is possible in principle to use their formal results to determine an exact percent bias reduction for a particular strategy under specific assumptions about the individual growth model. These formulas require, however, specification of a functional form for individual growth, and parameters of the growth curve distributions for each group. Since these are not known in the typical premeasure/postmeasure study, these formulas have little practical utility. Further, if the information were available, we probably would not use an adjustment strategy, but rather a procedure based on the growth curve methodology discussed later in Section 12.7.

We are left at this point with one general conclusion. The routine application of adjustment methods, based on the variation/covariation in the pre- and postmeasures, does not guarantee effective bias reduction. While for simplicity we have focused here on linear adjustment methods, it can be shown that matching and standardization methods yield similar results. Thus our conclusion applies with equal force both to gain score strategies and to all methods introduced in Chapters 6 to 10.

12.6 DATA ANALYTIC ADVICE

The individual growth trajectory in the absence of an intervention often represents a major confounding variable in premeasure/postmeasure studies. While the premeasure may be highly correlated with the slope of the growth trajectory, the premeasure will not contain full information about it. Thus we have a confounding variable, the individual trajectory, part of which, the slope, remains unmeasured. As discussed in Chapter 5, such an unmeasured confounding variable can produce a biased estimate of treatment effects.

The bias that we seek to reduce is formed by a selection or assignment process which allocates subjects differentially to groups depending upon their individual growth characteristics. This process creates differences among the groups on the premeasure and postmeasure means even in the absence of a treatment. In order to analyze these data appropriately, we must identify a model that can adequately describe the growth *in the absence of any intervention.*

Unfortunately, the pre- and postmeasure alone provide insufficient data to identify such a model correctly. We must therefore look outside these data for information on the actual process of selecting subjects for the treatment and comparison groups. On this basis, we might, for example, begin to build some logical arguments favoring one of the four cases discussed above. For example, Campbell and Erlebacher in their critique of the Head Start evaluation (previously discussed in Section 12.3) built some persuasive arguments that this was a fan spread situation [case 3 of (12.9)]. Conditional on this assumption, one could then choose an analytic strategy for these data that generated an adjustment coefficient consistent with this model. Clearly, the validity of the estimated program effects is conditional upon the validity of the selection process assumptions.

This approach, however, will not work in some instances. For example, if the trajectories cross over, or if prior treatment effects are present, such as in the Head Start/Follow Through example, no linear adjustment method is likely to generate an appropriate program effect estimate. In these cases, the analyst may be forced to set an a priori value on either the adjustment coefficient or the actual amount of adjustment. This may seem quite arbitrary, yet a simple gains analysis does the same (i.e., it sets $\hat{\beta} = 1$). Again, substantive knowledge of the setting, in this case information on the selection and change process, seems essential if the adjustments are to be performed in an appropriate manner. A variety of adjustments should be tried, and the analyst should assess the sensitivity of the estimated treatment effects to these alternatives.

At the very least, when we analyze a premeasure/postmeasure data set, we should examine the resultant $\hat{\beta}$ to see what it implies about the form of the mean growth trajectories in the absence of the intervention. If we cannot justify this assumption on the basis of our knowledge about the growth and selection processes operant in this case, the results of the analysis should be viewed with caution.

In many applications, information exists on other background variables in addition to the premeasure. While the individual growth trajectories are still the major confounding factor in the study, these other variables can sometimes be usefully incorporated in the analysis. For example, in the HSPV study, ethnic membership was considered an important background variable. A simple gains analysis might use this discrete variable as a blocking factor, essentially performing separate gains analyses for each group. More generally, all the methods introduced in Chapters 6 to 10 can be employed with confounding variables in addition to the premeasure. While such analyses can be very effective in specific cases, they do not "solve the problem." If anything, these strategies may simply obfuscate the nature of confounding in the complexities of multivariate analysis.

For the present, the best advice we can offer is to let the heuristic framework of Sections 12.4 and 12.5 and knowledge of the specific context guide the analysis. No empirical strategy currently exists that allows us to remove the substantive and "data analytic skill" components from this effort. It should be noted that some recent work (Cain, 1975; Kenny, 1975; Rubin, 1977, 1978; Cronbach et al., 1977) has focused on building models of the selection process and gathering data on it. This can be used as a basis for developing new methods to estimate treatment effects in premeasure/postmeasure data sets. Although this holds promise for the future, the work to date has been limited to some very specialized cases.

12.7   NEW DIRECTIONS: DESIGN AND ANALYSIS STRATEGIES BASED ON INDIVIDUAL GROWTH CURVES

12.7.1   Individual Growth Curve Models

Another promising approach in the design and analysis of premeasure/postmeasure studies focuses on developing new methods based on individual growth curve models (Bryk and Weisberg, 1976; Strenio et al., 1977; Strenio et al., 1979). Traditional analysis methods use the adjusted postmeasure mean for the control group as the standard of comparison for assessing the effects of the treatment. In terms of linear adjustment methods, we use $\overline{Y}_0(t_2) + \hat{\beta}[\overline{Y}_1(t_1) - \overline{Y}_0(t_1)]$ to estimate the correct standard of comparison, $\overline{Y}_1^*(t_2)$. Analysis strategies based on individual growth curve models, however, develop their standard of comparison from a different perspective. Rather than using the adjusted mean of a comparison group, this alternative focuses on the natural growth of subjects prior to exposure to treatments, attempting to project explicitly a postmeasure status for the treatment group as if they had been subject to the control condition. The actual growth is then compared with projected growth, the difference representing the effects of the treatment.

The model assumes that over the duration between premeasure and postmeasure each individual's growth consists of two components: (*1*) systematic growth, which can be characterized by a growth rate and a corresponding growth curve; and (*b*) an individual noise or random component, which is specific to a particular subject at a certain point in time. Thus we can represent the observed score for individual i at any time t as

$$Y_i(t) = G_i(t) + R_i(t), \tag{12.10}$$

where $G_i(t)$ represents systematic growth and $R_i(t)$ represents the random component.

The individual's systematic growth, $G_i(t)$, is represented as a function of age (or some other time metric). While in principle this function may take any form, it may often be adequate to assume that it is linear:

$$G_i(t) = \pi_i a_i(t) + \delta_i, \tag{12.11}$$

where π_i represents the slope, δ_i represents the Y intercept, and $a_i(t)$ is the age for subject i at time t. Individuals may vary in terms of a growth rate, π, and an intercept parameter, δ. The model assumes that π and δ are distributed with means μ_π and μ_δ, variances σ_π^2 and σ_δ^2, and covariance $\sigma_{\pi\delta}$.

Note that this represents the simplest model for $G(t)$, which incorporates varying individual growth. While too simple to fully describe many growth processes, linear individual growth may be a reasonable analytic approximation over a short term even if long-term growth has a more complex form.

As for the random component, the model assumes that

$$E[R_i(t)] = 0$$

$$\text{Var}[R_i(t)] = \sigma_R^2 \qquad (12.12)$$

(i.e., fixed over subjects and time) and

$$\text{Cov}[R_i(t), \pi] = \text{Cov}[R_i(t), \delta] = 0.$$

Thus we represent the observed premeasure ($t = t_1$) as

$$Y_i(t_1) = \pi_i a_i(t_1) + \delta_i + R_i(t_1). \qquad (12.13)$$

For convenience, let us define

$$\Delta_i = a_i(t_2) - a_i(t_1), \qquad (12.14)$$

where Δ_i represents the time duration between pre- and postmeasure for subject i. Note that we are assuming that t_1 and t_2 may differ across subjects, but are dropping the subscript i for notational convenience.

At the postmeasure point ($t = t_2$), in the absence of a treatment, we would have

$$Y_i(t_2) = \pi_i a_i(t_2) + \delta_i + R_i(t_2)$$

$$= G_i(t_1) + \pi_i \Delta_i + R_i(t_2), \qquad (12.15)$$

where $\pi_i \Delta_i$ represents the expected growth between pre- and postmeasure due solely to natural maturation.

In the presence of a treatment, we assume that over the time interval t_1 to t_2 the treatment increases each subject's growth by an amount v_i. This increment, v_i, has been termed the value added by the treatment (Bryk and Weisberg, 1976). Thus we can represent the measured growth subject i achieves by time t_2 under an intervention as

$$Y_i(t_2) = G_i(t_1) + \pi_i \Delta_i + v_i + R_i(t_2). \qquad (12.16)$$

Under this model, the treatment effect is fully described by the distribution of the v_i. We assume that v is a random variable with mean μ_v, and variance σ_v^2. Normally, we are interested in a summary measure of the treatment effect. This suggests that we estimate μ_v, the average of the individual treatment effects. During the period between the pre- and postmeasure, the observed change in the treatment group is $\overline{Y}_1(t_2) - \overline{Y}_1(t_1)$. The expected growth under the model is $\mu_\pi \overline{\Delta}$. If we knew the value of μ_π, a natural estimator of μ_v would be

$$\hat{\mu}_v = \overline{Y}_1(t_2) - \overline{Y}_1(t_1) - \mu_\pi \overline{\Delta}. \qquad (12.17)$$

Bryk et al. (1980) have shown, under the assumption that π and δ are independent of $a(t_1)$ and that π and Δ are independent, that the ordinary least squares regression of $Y(t_1)$ on $a(t_1)$ yields an unbiased estimate of μ_π, and as a result,

$$V = \overline{Y}_1(t_2) - \overline{Y}_1(t_1) - \hat{\mu}_\pi \overline{\Delta} \tag{12.18}$$

represents an unbiased estimate of μ_v.

12.7.2 An Illustration

We illustrate the technique using a different set of data from the Head Start study. The outcome measures are again child scores on the Pre-School Inventory. We have the following information:

$$n = 97$$
$$\overline{Y}_1(t_1) = 14.12$$
$$\overline{Y}_1(t_2) = 20.45$$
$$\overline{\Delta} = 7.40 \text{ months.}$$

Figure 12.9 presents a scatterplot of the premeasure scores on age. The ordinary

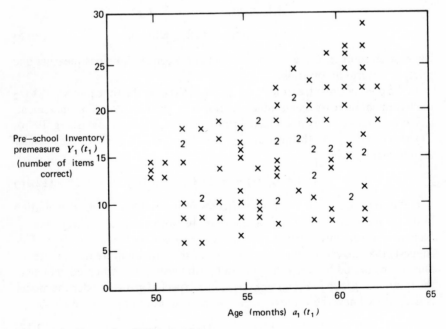

Figure 12.9   Scatterplot of PSI pretest by age.

least squares regression of $Y(t_1)$ on $a(t_1)$ yields

$$\hat{Y}_1(t_1) = -13.9 + 0.48[a(t_1)],$$

and we have $\hat{\mu}_\pi = 0.48$. From (12.18), the estimate of the treatment effect, μ_v, is given by

$$V = 20.45 - 14.12 - .48(7.40) = 2.75 \text{ points.}$$

While Bryk et al. (1980) do not derive an estimate of the standard error of V, they suggest the use of the jackknife technique (described in Chapter 8 of Mosteller and Tukey, 1977) to provide both a test statistic and standard error of V. Using the jackknife on the Head Start data resulted in a test statistic of 2.76 and a standard error of 1.19. The resultant t value of 2.32 with 96 degrees of freedom is significant beyond the .05 level.

12.7.3 Analysis of Model Assumptions

The technique can generate an unbiased estimate of a treatment effect conditional upon some important assumptions. The method as illustrated above uses the cross-sectional relationship between status and age at a particular point in time, t_1, to estimate the mean growth rate for the program group. This approach requires the important assumption that individual growth characteristics (reflected by π_i and δ_i in our model) are independent of age.

Nonindependence can occur in at least two different ways. First, in the population from which individuals are sampled, there may be historical trends causing age-related differences among the cases. For example, if there had been a major outbreak of rubella, the children conceived during that period might be developing at different rates than their counterparts who are either slightly younger or older. Second, even if this stable universe assumption is true for the population being studied, the process of selecting the sample may introduce an age by growth parameter relationship. For example, in analyzing the Planned Variation Head Start data, we were initially concerned that both the oldest and youngest children might be somewhat different from the rest. It was hypothesized that the oldest might be a delayed entry into the school group (i.e., slower growth rates) and the youngest perhaps somewhat more precocious than average (i.e., faster growth rates).

Another possible problem is that individual growth may be nonlinear. With extreme nonlinearity, the linear approximations will not be trustworthy even in the short term. If the analyst has reasons to suspect such nonlinearity, a transformation of the outcome [e.g., $\ln Y(t)$] may suffice. Finally, we should note that while the growth curve strategy does not require data on a comparison group, if one is available it could prove useful in the process of identifying an appropriate growth model.

12.7.4   Further Developments with Individual Growth Curve Models

The basic model can be extended in a number of directions. One is the incorporation of background variables (e.g., information on sex, social class, ethnic membership, etc.) into the model. A natural strategy here is to assume that the individual growth rate, π_i, is a linear function of the background variables:

$$\pi_i = \beta_0 + \sum_{j=1}^{J} \beta_j X_{ij} + \epsilon_i \qquad j = 1, \ldots, J, \qquad (12.19)$$

where X_{ij} is the value on variable j for subject i. Bryk et al. (1980) illustrate how the basic estimation procedure introduced above can be applied in this context. The assumptions discussed in Section 12.7.3, however, are still required.

Perhaps the most promising extensions of the model are to designs involving multiple premeasure points. Regardless of whether we have comparison group data or not, the validity of results from a single premeasure/postmeasure design will remain questionable for reasons discussed in previous sections. In the case of strategies based on individual growth models, we have gone quite far with only two cross sections of data. The first, the premeasure, we use as a proxy for longitudinal information to project future status. The second, the postmeasure, we use to assess actual progress. Clearly, the major weakness is in the first area. The next logical step is designs with two or more premeasure points on each subject. Through combination of the cross-sectional data and the longitudinal trajectories on each case, we should be able to estimate more precisely both the mean effect μ_v and the other aspects of the distribution of individual effects. Recent work (Fearn, 1975; Strenio et al., 1979) are first steps in this direction.

12.8   SUMMARY AND CONCLUSIONS

In the course of this chapter we have outlined a number of pitfalls in the analysis of data from premeasure/postmeasure designs. We have focused primarily on how one uses the premeasure information. We wish to remind the reader of our earlier discussion in Chapter 5 about the problems resulting from errors of measurement in the independent variables. In many applications, particularly in education and psychology, these errors can be quite substantial. In these instances, they further limit the bias reduction achieved with techniques such as ANCOVA.

While analyses of premeasure/postmeasure studies may appear quite precarious, such designs do represent a qualitative improvement over nonrandomized studies lacking premeasure information. These data often provide some sense of the likely magnitude and direction of bias in the posttreatment measures.

The one point we wish to emphasize strongly, however, is that the blind application of the statistical techniques cannot guarantee valid inferences. While such data can be effectively analyzed, they require considerable care.

Similarly, considerable care should be exercised in the process of study design and data collection. It is unfortunate that many past efforts with premeasure/postmeasure designs have been very haphazard in this regard. Although random assignment may not be possible, effort should be expended to equate the treatment and control groups on variables thought to be strongly related to outcomes. Where this is not possible, the investigator should attempt to gather as much information as possible on the actual selection process, and perhaps even attempt to build empirical models for it. Such information would prove invaluable in helping either to select an appropriate method or to develop one of the alternative approaches discussed in the last section. In short, we should try to minimize the burden placed on the statistical adjustment, while maximizing our reliance on understanding and knowledge of the particular research data.

Finally, a few words of caution are offered about the models suggested in the last section. While they may be helpful in some circumstances, they do not constitute a simple answer. Each requires a number of assumptions that in general are untestable. While, in principle, they compensate for many of the problems raised earlier, their ability to achieve this depends entirely on the legitimacy of the assumptions in each case. A wise approach in many cases is to consider multiple analysis strategies. The data analysis should not be limited to a single preconceived plan. It should be firmly rooted in exploratory analysis to examine what the data suggest about possible biases. This should be followed by alternative analyses, perhaps based on several of the methods described above. Finally, sensitivity analyses associated with each method should be conducted where feasible to ascertain the range of treatment effect estimates possible.

REFERENCES

Blomquist, N. (1977), On the Relation between Change and Initial Value, *Journal of the American Statistical Association,* **72**(360), 746-749.

Bryk, A. S. (1977), An Investigation of the Effectiveness of Alternative Adjustment Strategies in the Analysis of Quasi-experimental Growth Data, unpublished doctoral dissertation, Harvard University.

Bryk, A. S., and Weisberg, H. I. (1976), Value-Added Analysis: A Dynamic Approach to the Estimation of Treatment Effects, *Journal of Educational Statistics,* **1**(2), 127-55.

Bryk, A. S., and Weisberg, H. I. (1977), Use of the Nonequivalent Control Group Design When Subjects Are Growing, *Psychological Bulletin,* **84**(5), 950-962.

Bryk, A. S., Strenio, J. F., and Weisberg, H. I. (1980), A Method for Estimating Treatment Effects When Individuals Are Growing, to appear in *Journal of Educational Statistics.*

Cain, G. G. (1975), Regression and Selection Models to Improve Nonexperimental Comparisons, *in* C. A. Bennett and A. A. Lumsdaine, Eds., *Evaluation and Experiment,* New York: Academic Press.

Campbell, D. T., and Erlebacher, A. (1970), How Regression Artifacts in Quasi-experimental Evaluations Can Mistakenly Make Compensatory Education Look Harmful, *in* J. Hellmuth, Ed., *Compensatory Education: A National Debate,* Vol. 3, *Disavantaged Child,* New York: Brunner-Mazel.

Cronbach, L. J., Rogosa, D. R., Floden, R. E., and Price, G. G. (1977), Analysis of Covariance in Nonrandomized Experiments: Parameters Effecting Bias, Stanford, CA: Stanford Evaluation Consortium.

Fearn, R. (1975), A Bayesian Approach to Growth Curves, *Biometrika,* **62**(1), 89–100.

Harris, C. W., Ed. (1967), *Problems in Measuring Change,* Madison, WI: University of Wisconsin Press.

Kenny, D. (1975), A Quasi-experimental Approach to Assessing Treatment Effects in the Nonequivalent Control Group Design, *Psychological Bulletin,* **82**(3), 345–362.

Linn, R. L., and Slinde, J. A. (1977), The Determination of Significance of Change between Pre- and Post-testing Periods, *Review of Educational Research,* **47**, 212–250.

Mosteller, F., and Tukey, J. W. (1977), *Data Analysis and Regression,* Reading, MA: Addison-Wesley.

Reichardt, C. S. (1979), The Statistical Analysis of Data from the Nonequivalent Control Group Design, *in* T. D. Cook and D. T. Campbell, Eds., *The Design and Analysis of Quasi-experiments in Field Settings,* Chicago: Rand McNally.

Rubin, D. B. (1977), Assignment to Treatment Groups on the Basis of a Covariate, *Journal of Educational Statistics,* **2**, 1–26.

Rubin, D. B. (1978), Bayesian Inference for Causal Effects: The Role of Randomization, *The Annals of Statistics,* **6**, 34–58.

Smith, M. S. (1973), Some Short Term Effects of Project Head Start: A Report of the Third Year of Planned Variation—1971–1972. Cambridge, MA: Huron Institute.

Strenio, J., Bryk, A. S., and Weisberg, H. I. (1977), An Individual Growth Model Perspective for Evaluating Educational Programs, *Proceedings of the 1977 American Statistical Association Meetings (Social Science Section).*

Strenio, J. F., Weisberg, H. I., and Bryk, A. S. (1979), Empirical Bayes Estimation of Individual Growth Curve Parameters and Their Relationship to Covariates, paper supported by Grant NIE-G-76-0090 from the National Institute of Education, U.S. Department of Health, Education and Welfare, Cambridge, Mass: The Huron Institute.

Winer, B. J. (1971), *Statistical Principles in Experimental Design,* 2nd ed., New York: McGraw-Hill.

Weisberg, H. I. (1973), Short-Term Cognitive Effects of Head Start Programs: A Report on the Third Year of Planned Variation—1971–1972, Cambridge, MA: Huron Institute.

Weisberg, H. I. (1979), Statistical Adjustments and Uncontrolled Studies, *Psychological Bulletin,* **86**, 1149–1164.

CHAPTER 13

Choice of Procedure

Given the procedures described in previous chapters, it is natural to ask which procedure is appropriate for a given situation. Unfortunately, this question has no simple answer. At the very least, the answer will depend on the number and type of variables, the population distributions of these variables, the quantity of data that are or can be collected, and whether the outcome variable (or risk variable in case-control studies) is already recorded or not.

For simplicity, we shall discuss this question for the special situation of a single confounding variable. Table 13.1 presents the eight cases that can arise, depending on whether the outcome variable, the risk variable, and confounding variable are categorical or numerical. For each case the appropriate adjustment procedures are listed along with applicable matching techniques.

With the exception of regression, all the procedures listed in Table 13.1 have been discussed in this book. In this chapter we provide guides to aid the inves-

Table 13.1   Summary of Appropriate Adjustment and Matching Procedures

	Form of Variables[a]				
Case	Outcome Variable	Risk Variable	Confounding Variable	Adjustment Procedure	Matching[b]
I	C	C	C	Standardization, stratification, log-linear, logit	Type 1
II	C	C	N	Logit	Type 2
III	N	C	C	Standardization, stratification, ANCOVA (cohort only), logit (case-control only)	Type 1, (cohort only)
IV	N	C	N	ANCOVA (cohort only), logit (case-control only)	Type 2, (cohort only)
V	C	N	C	Logit (cohort only)	Type 1, (case-control only)
VI	C	N	N	Logit (cohort only)	Type 2, (case-control only)
VII	N	N	C	Regression	None
VIII	N	N	N	Regression	None

[a] C, categorical; N, numerical.

[b] Type 1: frequency or stratified matching; type 2: mean, frequency, stratified, caliper, or nearest neighbor matching.

tigator in choosing among these procedures. These guides, discussed in the following three sections, fall into three categories:

1. Whether to treat a given variable as categorical or numerical.
2. Whether to match or adjust.
3. Whether to use a combination of procedures.

13.1   CATEGORICAL OR NUMERICAL VARIABLES

Although a variable may be recorded as numerical, the experimenter always has the choice of changing it to a categorical variable by stratification. The standardization example in Section 7.1 is an example of such a change; we replaced the numerical variable age by a series of age categories "15–34," "35–44," and so on. This is known as stratifying age. Similarly, a categorical variable whose categories are ordered, for example nonsmoker, light smoker, and heavy smoker, can be treated as categorical or the categories may be scored numerically: nonsmoker = 1, light smoker = 2, and heavy smoker = 3. Therefore, although we may really be in Case II of Table 13.1, for example, we may opt for treating the numerical confounding variable as a categorical one, and so use methods appropriate for Case I.

This choice leads to two questions: Why would we change the form of a variable? What are the effects of such a change? It is usually undesirable to change the form of the outcome variable, since the validity of the underlying statistical model will be affected. Accordingly, we shall discuss changing the form of the risk and confounding variables only.

13.1.1 Why Change the Form of a Variable?

"Why?" is the easier question. The investigator may be more familiar with the methods for analyzing numerical variables or may only have access to computer programs for doing so. On the other hand, appropriate analysis of a numerical variable may require too stringent assumptions, such as linearity of the response relationship. The analysis of a numerical variable made categorical by stratification no longer requires linearity. For example, it is known that cancer mortality is related to age, but the form of the relation is not known. By stratifying age, as done in Section 7.1, we avoid the necessity of specifying the relationship of mortality to age.

13.1.2 What Is the Effect of a Change in the Form of a Variable?

The effect of changing the form of the variable is the more important statistical question. Unfortunately, very little research has been conducted on this topic. Treating the stratification of numerical variables first, it is clear that the categorical variable is only an approximation to the underlying numerical one. Therefore, some penalty must be paid since information is being lost. Two important investigations concerning the size of the penalty are those of Cochran (1968) and McKinlay (1975).

Case IV to Case III. Cochran (1968) considers the consequences of making a numerical confounding variable categorical, that is, moving from Case IV (categorical risk variable and numerical confounding and outcome variables) to Case III (categorical risk and confounding variables and numerical outcome variables). Specifically, he considers the effectiveness of stratification in removing the bias in the estimators of the difference in means, as compared to analysis of covariance (ANCOVA).

Cochran assumes that the regression of the response (Y) on the confounding variable (X) is linear with identical slopes in the two risk factor groups, but that the distribution of X is different in the two groups. If the response relationship is nonlinear, Cochran then assumes that the variable, X, represents an appropriate transformation of the confounding variable so that the regression is linear in X. Next, he stratifies the distribution of X into K strata, and computes the bias before and after stratification for various choices of distributions for X and for various values of K. He finds that stratifying the population distribution of the numerical confounding variable of one of the groups being compared into six strata of equal numbers of subjects generally leads to a removal of 90 to 95% of the bias. Cochran found these percentages to be quite consistent over the various choices of distributions for X he investigated—normal, chi-square, t, and beta distributions—and even for a nonlinear regression of Y which is approximately a cubic function of X. These results are based on a known population distribution; however, similar results would be expected for stratification based on a large sample distribution.

Cochran's results can be interpreted as follows. Although for a parallel linear response function, ANCOVA is the appropriate adjustment procedure for Case IV of Table 13.1, and, as such, is preferable to stratification, stratifying the confounding variable is an alternative method that performs adequately under a variety of situations. Regardless of the form of the regression of Y on X, as long as the regressions in the two risk factor groups are parallel, the experimenter can expect to remove 90 to 95% of the bias when stratifying the confounding variable into six strata of equal numbers of subjects.

Cochran's results also indicate that the percentage of bias removed critically depends on the number of strata used. For a variety of frequently encountered distributions of the confounding variable, Cochran found that with five strata with equal numbers of subjects, the percentage bias reduction is in excess of 88%, with four strata in excess of 84%, with three in excess of 77%, and with two strata in excess of 60%. (See Table 6.5 for the exact percentages based on a normally distributed confounding variable.) Five or six strata would appear to be the minimum number necessary, and little extra reduction is gained by having more than six strata. Cochran points out that these percentages also apply approximately if X is an ordered categorical variable representing a grouping of an underlying numerical variable. We will return to this study in Section 13.2.2.

Case II to Case I.   McKinlay (1975) looks at the effect of moving from Case II to Case I (i.e., stratifying a numerical confounding variable). With the outcome variable categorical and the confounding variable numerical, the adjustment procedure to use is logit analysis, which yields an estimate of the odds ratio. As discussed in Chapter 9, the logit estimator of the odds ratio is biased, although this bias will be small when the sample size is large. Therefore, in contrast to ANCOVA examined by Cochran, logit analysis will not remove 100% of the bias, so it is more difficult to judge the effectiveness of the procedures studied by McKinlay for analyzing Case II as a Case I type of situation.

The major part of McKinlay's work is a Monte Carlo study of the mean squared errors* of the modified Woolf, Mantel–Haenszel, and Birch estimators as applied to stratified samples versus the crude estimator, which ignores the confounding variable by not stratifying (the stratified estimators are given in Section 7.6). McKinlay considered up to four strata. With few exceptions, primarily for the Birch estimator, the mean squared errors of the stratified procedures were substantially lower than for the crude estimator. However, only the Mantel–Haenszel estimator consistently removed bias in all the cases considered. For the other estimators, the bias was actually increased in some cases, but this was masked in the mean squared error. McKinlay concludes that the modified Woolf estimator is best, particularly when the bias in the crude odds ratio estimator, before stratification, is not large. Unfortunately, McKinlay did not obtain results for the logit analysis estimator. We will also discuss this study in Section 13.2.2.

Numerical Scoring of Categorical Variables. Treating an ordered categorical variable as numerical (e.g., low = 1, medium = 2, and high = 4) presents a more difficult problem, since it is usually not clear exactly how to recode such a variable. For the categorical variable with only two categories, this is not a problem, since all choices of recoding are equivalent. However, for more than two categories, we must select coded values that meaningfully reflect relative magnitudes, so we must know something about the variable in question. Frequently, the ordered categorical variable reflects some underlying numerical variable that could not be or was not measured. Recoding such a variable requires identifying this underlying numerical variable or at least some approximation to it. There may be little justification for arbitrarily recoding "low–medium–high," for example, as "1–2–3." Recoding as "1–2–3" should only be done if the difference between a "high" and a "medium" value is judged approximately equal to the difference between a "medium" and a "low." If, for example, the difference between "high" and "medium" is judged to be roughly twice the difference between "medium" and "low," it would be better to recode the values as "1–2–4." If age has been recorded in intervals, yielding an ordered categorical variable, we could recode the ordered variable as a numerical one by using the interval midpoints.

The effect on the bias in the estimated treatment effect of using an ordered categorical variable as numerical has not been studied. However, because the application of ANCOVA to such a variable is a special case of regression with grouped data, we can refer to Haitovsky (1973). Haitovsky showed that using

---

* The mean squared error equals the variance plus the bias squared. For a derivation of this result, see Appendix 2A.

information on grouped data as an approximation for the underlying unobservable data will lead to biased estimates of the slope, β. As ANCOVA depends on an unbiased estimate of β, we infer that ANCOVA on grouped data will yield biased estimates of the treatment effect, $\alpha_1 - \alpha_0$.

13.2 COMPARISON OF MATCHING AND ADJUSTMENT PROCEDURES

In comparing matching with an adjustment procedure, such as ANCOVA, we are faced with the problem of unequal sample sizes. As pointed out by McKinlay (1974), often it would be more realistic to account for the loss of the unmatched units when comparing matching with adjustment techniques. This issue is important in answering the question: Should we match or use an adjustment procedure? We return to this topic in Section 13.2.3. In Section 13.2.1 and in the beginning of Section 13.2.2, however, we will first assume that there are no losses due to matching. In these latter two sections, we will compare matching versus ANCOVA and matching versus stratification.

13.2.1 Matching versus Analysis of Covariance

In this section we compare matching methods and analysis of covariance in estimating the treatment effect for Case IV studies (see Table 13.1), where both the response and confounding variable are numerical and the risk variable is categorical. Matching methods and ANCOVA have been compared both from an analytic point of view, Cochran (1953), and via Monte Carlo studies, Billewicz (1965) and Rubin (1973a, b). For a summary and some extensions of Rubin's work, see Cochran and Rubin (1973). The specific matching techniques considered are primarily various pair matching methods, with some references to frequency matching (see Section 6.7).

In the Monte Carlo studies the values of the confounding variable are simulated from distributions approximating "real-life" situations and an effort is made to compare the different methods over a wide range of conditions. Rubin concentrated on an evaluation of the bias reduction obtained by the different techniques, whereas Billewicz investigated the difference in precision with which the treatment effect is estimated.

In the literature we encounter comparisons between matching, ANCOVA on random samples, ANCOVA on matched samples ignoring the matching, and regression adjustment on matched samples. (Regression adjustment on matched samples is presented in Section 13.3.1.) In our discussion we will assume a cohort study and that the relations between the outcome and confounding variables in the two treatment groups are parallel. The linear and nonlinear cases are considered separately.

Linear Regression. In the case of linear parallel regressions, ANCOVA and regression adjustment on matched samples yield unbiased estimates of the treatment effect. Since exact matching of numerical variables is not possible, any matching method will yield biased estimates although the bias may be small if the matches are close.

In addition, Billewicz (1965) found that stratified matching using three categories was less precise than ANCOVA or ANCOVA on frequency-matched samples. Carpenter (1977) considered a multivariate case where close matches are likely. He found ANCOVA on circular matched samples to be more precise than either circular matching or ANCOVA alone. Whether ANCOVA was more precise than circular matching alone depended on the number of confounding variables and the control reservoir size. Fewer variables and larger reservoir sizes favored matching.

On these bases, ANCOVA on random or matched samples is preferred to matching alone in the linear parallel case. However, this presupposes that the response relations are known to be linear and parallel, an assumption that is rarely valid in practice. We emphasize again that neither Cochran, Billewicz, nor Carpenter considered the potential loss of sample size due to matching.

Nonlinear Regression. With a nonlinear but parallel relationship between the confounding and outcome variables, one might intuitively expect that pair matching would be superior to ANCOVA. The investigator might reason that it is better to match subjects and make no assumptions as to the exact nature of the model than to rely on analysis of covariance and possibly use an inappropriate model to eliminate the effect of the confounding variable.

Empirical examination of this comparison has been conducted only for the case where the relationship is moderately nonlinear in the range of interest. [As mentioned in Section 6.10.5, an extreme example of what might be termed moderately nonlinear is $\exp(X)$, and a more reasonable example would be $\exp(X/2)$.] This is intended to exclude the violent nonlinearity that an alert investigator should be able to detect and realize that ANCOVA is not appropriate.

Both Rubin (1973b) and Billewicz (1965) came to the conclusion that ANCOVA was preferable to matching. Rubin found that ANCOVA was more effective in removing bias than was random-order nearest available matching. But he also found that regression adjustment on nearest available matched samples was generally superior to ANCOVA, although the difference was usually small. Similarly, Rubin (1979) found that the same general conclusion held for multivariate metric matching (see Section 6.10.3). On the basis of precision, Billewicz also preferred ANCOVA to matching (stratified matching with three strata), and he found that ANCOVA on frequency-matched samples resulted in the same precision as ANCOVA.

In conclusion, for moderately nonlinear response relationships, ANCOVA or matching followed by ANCOVA would be preferred to matching alone.

13.2.2 Matching versus Stratification

In comparing matching with stratification, we will distinguish two situations: the categorical and the numerical confounding variables. In both situations, we bring to bear the earlier discussion in Section 13.1 of the effect of changing the form of the confounding variable, more specifically, the effect of treating a numerical confounding variable as categorical.

Categorical Confounding Variable.   Because all categories are exact for a categorical confounding variable, all bias due to such a confounding variable can be removed in a comparison of responses, at least in large samples, whether through stratification or matching. Therefore, the choice between matching and stratification in large samples must be based on precision. There are two considerations. First, matching pays a penalty because of the possible reduction of the sample size. Conversely, the stratified estimator will be less precise than an estimator based on frequency-matched samples of the same size, particularly if the numbers in some of the strata are small (Cochran, 1965). Consequently, in large samples we favor matching for categorical confounding variables, if there would be little or no reduction in sample size. Otherwise, stratification is to be preferred.

For estimation of the odds ratio from small samples where the estimates are biased, further research is needed before firm recommendations can be made. McKinlay's (1975) results, reported below, do suggest, however, that stratification may be preferable.

Numerical Confounding Variable.   In many situations the confounding variable is numerical, or a categorical confounding variable has so many values that categories are collapsed before stratification can be used. In these situations, bias is never completely removed. Therefore, the relative effectiveness of matching and stratification in removing bias is an issue in these cases.

McKinlay (1975) investigated the properties of estimators of the odds ratio for a numerical confounding variable and a dichotomous outcome variable. In this Monte Carlo study, first described in Section 13.1.2, she compared the bias, variance, and mean squared error of the odds ratio estimators from unstratified, stratified (with several estimators), and stratified matching methods. Of particular importance is that she explicitly evaluated the effect of reduced sample size when forming matched pairs.

McKinlay's (1975) results showed that the matched estimator was less desirable than the stratified estimators. The matched pair estimator was subject to an inflated bias and a larger variance relative to other estimators. McKinlay concluded that the sample losses and the extra costs incurred through pair matching (in terms of time needed to form matches) cannot be justified.

For a numerical outcome variable we again refer to analytic and Monte Carlo results given in Cochran (1968), discussed in part in Section 13.1.2, that compare frequency matching with stratification. Cochran found that if, after stratification, the within-strata distributions of the confounding variable do not greatly differ from the treatment and comparison groups, frequency matching and stratification will be almost free of bias (due to the confounding variable). Indeed, results based on a normally distributed confounding variable divided into six strata show that more than 90% of the bias would be removed by either method.

Based on Cochran's results, we conclude that there is not much difference in the percentage bias reductions between frequency matching and stratification, with frequency matching favored because of slightly more precise estimators in small samples. In this comparison Cochran did not take into account the potential reduction in sample size due to matching.

13.2.3 Should We Match?

All the foregoing discussion comparing matching to an adjustment procedure still leaves unanswered the question of when should one match. An important consideration in answering this question is whether or not matching will ultimately result in a smaller sample size being used.

In conducting a comparative study, two situations can be recognized. In the first, we have a reservoir of subjects from each of two groups and have recorded the measurements on all relevant variables, including the outcome variable in cohort studies or the risk variable in case-control studies. In the second, we again have a reservoir of subjects and measurements on the confounding variables, but we have not yet measured the outcome variable (cohort study) or risk variable (case-control study), possibly because this measurement is expensive or time-consuming. (We may have to wait until something has happened.) In the first situation, matching as compared to ANCOVA or stratification will generally result in a reduction of final sample size as the number of matched pairs can be no larger than the size of the smallest reservoir. Therefore, matching should be compared to an adjustment procedure that uses a larger sample. McKinlay's (1975) results indicate that if this is done, matching will come in a poor second. The conclusion is not to match if it means throwing away a large fraction of the data.

If recording the value for the outcome variable (or risk variable in case-control studies) is very time-consuming or expensive, we may only want to record such information for a limited number of units, and the issue then is how to select these units in order to obtain the best estimate of the treatment effect. Indeed, matching is a method to be recommended for such a situation. Assuming that the reservoir is large enough that the desired number of pairs can be found, matching is a particularly good choice in small samples because of the precision with which the treatment effect can still be estimated. If the sample size is too small, adjustment procedures cannot be used with any accuracy. For example, if one attempts to use stratification on small samples, there is a good chance that many of the strata will be empty for one or the other group. Frequency matching, on the other hand, would ensure members of both groups within each stratum (see Cochran, 1953, and Fleiss, 1973). The precision advantages of matching can be substantial; Carpenter (1977) found that under reasonable assumptions, matching (specifically circular matching) can double the precision of the estimated treatment effect as compared to ANCOVA for the same sample sizes.

The problem of reduced sample size led Rubin to develop the nearest available matching techniques (Section 6.5). When matching on a numerical confounding variable, the nearest available matching procedure can be used with no loss of sample size. Unfortunately, the relative merits of this matching technique versus stratification have not been studied. Nearest available matching does well in comparison to ANCOVA, so we can infer that this method will compare more favorably to stratification than did the stratified matching technique considered by McKinlay (1975).

Matching does have one clear advantage over such techniques as ANCOVA and logit analysis, which assume a particular functional form for the relation between the outcome and confounding variables. With the exception of mean matching, matching (and stratification) makes no assumptions about the functional form, and as such the researcher is protected against model specification errors. For reasonably large samples and if there is no loss of sample size due to matching, for example, by using nearest available matching, matching compares favorable to ANCOVA. If the researcher is unwilling to assume linearity, matching would certainly be a good choice.

13.3 COMBINING PROCEDURES

In many cases it is desirable to apply more than one bias-reduction procedure to the same data. The larger and more complex the data, the more likely this will be, as it will be difficult to find one procedure that can properly handle all types of confounding variables. For example, the more confounding variables one must deal with, the harder it will be to assume that the outcome variable or some transformation is linearly related to all the confounding variables, a necessary assumption for ANCOVA.

There are basically two ways that we can apply two procedures to the same data. First, apply the two procedures to the same confounding variable; second, apply the two procedures to different confounding variables. An example of the first is log-linear analysis prior to standarization. As discussed in Chapter 7, this increases the precision of directly standardized rates. Another example, discussed in Section 13.2, is regression adjustment on pair-matched samples as a combination of ANCOVA and matching.

Although other combinations are possible, the discussion of using different procedures for different confounding variables will be restricted to the problem of applying adjustment procedures to pair-matched samples. This is both a useful and a nontrivial case—useful because additional confounding variables that arise only after the sample is chosen obviously cannot be accounted for in the matching process by which the samples were chosen; nontrivial because matching introduces dependence, so the adjustment procedures must take the dependence into account.

13.3.1 Regression Analysis of Pair-Matched Samples

For a numerical outcome let us define Y_{ij} as the response for the ith member of the jth matched pair and X as a confounding variable that was not taken into account in the matching process. The treatment group is represented by $i = 1$, and the comparison group by $i = 0$. The ANCOVA model is

$$Y_{ij} = \alpha_i + \beta X_{ij} + e_{ij} \qquad i = 0, 1; j = 1, 2, \ldots, N, \qquad (13.1)$$

where e_{ij} includes the effect of the matching variable. Because of the matching, e_{1j} and e_{0j} are not independent, and thus the ANCOVA assumptions are violated. The standard approach to circumvent the problem, as is done for example with the paired-t test, is to work with the differences:

$$Y_{1j} - Y_{0j} = (\alpha_1 - \alpha_0) + \beta(X_{1j} - X_{0j}) + (e_{1j} - e_{0j}) \qquad j = 1, 2, \ldots, N. \qquad (13.2)$$

The regression of $Y_{1j} - Y_{0j}$ on $X_{1j} - X_{0j}$ then yields an unbiased estimator of $\alpha_1 - \alpha_0$. Rosner and Hennekens (1978) give a hypothetical example of the application of this procedure.

Rubin (1970) motivates this procedure differently. He assumes that the regression of Y on X is nonlinear, but parallel in the two groups, and that the nonlinear terms in the regression are dominated by the linear terms. He then shows that the procedure of matching on X and then estimating $\alpha_1 - \alpha_0$ based on (13.2) is less affected by the nonlinearity than is ANCOVA. This is the procedure termed regression adjustment on matched samples that has been discussed earlier (Section 13.2.1). Note that in Rubin's situation, the matching and regression adjustment are performed on the *same* confounding variable.

13.3.2 Logit Analysis of Pair-Matched Samples

For a dichotomous response variable, logit analysis serves the same function as ANCOVA does for numerical responses. The logit model discussed in Chapter 9 can be generalized as, see Cox (1970, p 56)

$$\ln \left[\frac{P(R, X, j)}{1 - P(R, X, j)} \right] = \alpha + \gamma R + \beta X_{Rj} + \delta_j, \qquad (13.3)$$

where R ($= 0$ or 1) is the risk variable; γ is the logarithm of the odds ratio, the parameter of interest; δ_j is a parameter that reflects the effect of the jth pair; and $P(R, X, j)$ is the probability of a "successful" outcome given R, X and j. (The meaning of δ_j will be further discussed below.) The parameter β indicates the effect of the confounding variable not taken into account during the matching, and X_{Rj} is the value of the confounding variable in the jth pair with risk factor value R. Straightforward application of logit analysis to this model is not appropriate because the number of parameters increases as the number of observations (pairs) grows. This violates an assumption required for the large-sample properties of maximum likelihood estimators to hold.

With a binary outcome variable with values denoted 0 and 1, the possible responses on a pair of cases are then 00, 10, 01, and 11, writing the response of the comparison group first and of the treatment group second. Since the responses 00 and 11 do not affect the comparison between the two levels of the risk variable, we have only to consider the discordant pairs (i.e., those pairs for which the response is 01 or 10).

The probability, P_j, of a 01 response for the jth pair, given that the pair is discordant, satisfies

$$\ln \left(\frac{P_j}{1 - P_j} \right) = \gamma + \beta(X_{1j} - X_{0j}). \qquad (13.4)$$

The logarithm of the odds ratio γ can then be estimated by fitting the logit model (13.4) in the usual way by treating a 01 pair as $Y = 1$, a 10 pair as $Y = 0$, and ignoring all 00 and 11 pairs. The procedure based on (13.4) is referred to as the conditional likelihood method (Cox and Hinkley, 1974), as it only analyzes the discordant pairs. Holford et al. (1978) and Rosner and Hennekens (1978) discuss the procedure further and give an example, and Breslow et al. (1978) consider the extension to more than one matched control.

The usual analysis of matched pairs when the response is dichotomous (see Chapter 6) is to estimate γ by

$$\hat{\gamma} = \ln \left(\frac{\text{number of 01 pairs}}{\text{number of 10 pairs}} \right). \qquad (13.5)$$

From (13.4) it follows that if we have perfect matching on X ($X_{1j} = X_{0j}$, for all j) or if β equals zero, the logit estimator of γ equals $\hat{\gamma}$ of (13.5).

It is sometimes possible to analyze matched samples as if they were not matched. The idea is to introduce the matching explicitly into the model. In the case of pairing on the basis of a set of confounding variables, the δ_j of (13.3) may be replaced by a set of parameters corresponding to the matching variables. For example, in the case of stratified matching on age and sex, the δ_j could be replaced by parameters for each of the age and sex categories and, to allow for age–sex interactions, various combinations of these variables. As the number of parameters in the model, although possibly large, would then not change as the number of pairs increases, maximum likelihood could be applied in the usual way and the conditional likelihood approach would not be necessary. An example of the use of this procedure is Smith et al. (1975).

However, this alternative method would not work, for example, in the case of natural pairing, such as in the case of siblings, where there may not be an explicit set of confounding variables. Matching is intended to control for such variables as experience and inheritance, which are clearly relevant but not reducible to a few simple variables. Also, note that, unlike the conditional likelihood approach, the alternative method described above requires knowledge of the relationship between the response and the matching variables.

REFERENCES

Billewicz, W. Z. (1965), The Efficiency of Matched Samples: An Empirical Investigation, *Biometrics,* **21,** 623–644.

Breslow, N. E., Day, N. E., and Halvorsen, K. T. (1978), Estimation of Multiple Relative Risk Functions in Matched Case-Control Studies, *American Journal of Epidemiology,* **108,** 299–307.

Carpenter, R. G. (1977), Matching When Covariables Are Normally Distributed, *Biometrika,* **64,** 299–307.

Cochran, W. G. (1953), Matching in Analytical Studies, *American Journal of Public Health and the Nation's Health,* **43,** 684–691.

Cochran, W. G. (1965), The Planning of Observational Studies of Human Populations, *Journal of the Royal Statistical Society, Series A,* **128,** 234–265.

Cochran, W. G. (1968), The Effectiveness of Adjustment by Subclassification in Removing Bias in Observational Studies, *Biometrics,* **24,** 295–314.

Cochran, W. G., and Rubin, D. B. (1973), Controlling Bias in Observational Studies: A Review, *Sankhyā, Series A,* **35,** 417–446.

Cox, D. R. (1970), *The Analysis of Binary Data,* London: Methuen.

Cox, D. R., and Hinkley, D. V. (1974), *Theoretical Statistics,* London: Chapman & Hall.

Fleiss, T. L., (1973), *Statistical Methods for Rates and Proportions,* New York: Wiley.

Haitovsky, Y. (1973), *Regression Estimation from Grouped Observations,* New York: Hafner.

Holford, T. R., White, C., and Kelsey, J. L. (1978), Multivariate Analysis for Matched Case-Control Studies, *Americal Journal of Epidemiology,* **107,** 245–256.

McKinlay, S. M. (1974), The Expected Number of Matches and Its Variance for Matched-Pair Designs, *Journal of the Royal Statistical Society, Series C,* **23,** 372–383.

McKinlay, S. M. (1975), The Effect of Bias on Estimators in Relative Risk for Pair-Matched and Stratified Samples, *Journal of the American Statistical Association,* **70,** 859–864.

Rosner, B., and Hennekens, C. H. (1978), Analytic Methods in Matched Pair Epidemiological Studies, *International Journal of Epidemiology,* **7,** 367–372.

Rubin, D. B. (1970), The Use of Matched Sampling and Regression Adjustment in Observational Studies, unpublished Ph.D. dissertation, Harvard University.

Rubin, D. B. (1973a), Matching to Remove Bias in Observational Studies, *Biometrics,* **29,** 159–183.

Rubin, D. B. (1973b), The Use of Matched Sampling and Regression Adjustments to Remove Bias in Observational Studies, *Biometrics,* **29,** 185–203.

Rubin, D. B. (1979), Using Multivariate Matched Sampling and Regression Adjustment to Control Bias in Observational Studies, *Journal of the American Statistical Association,* **74,** 318–328.

Smith, D. C., Prentice, R., Thompson, D. J., and Herrmann, W. L. (1975), Association of Exogenous Estrogen and Endometrial Carcinoma, *The New England Journal of Medicine,* **293,** 1164–1167.

CHAPTER 14

Considerations in Assessing
Association and Causality

Up to now we have concentrated on methods which, when properly used, will provide relatively undistorted estimates of the treatment effect or association between two factors of interest. The phrase "treatment effect" will correctly designate this association if causality can be established. In this chapter the discussion emphasizes criteria which, when combined with informed judgment, can be used to assess causality from significant associations. Undistorted estimates of association are necessary if one is to accurately assess cause and effect. Tests of significance when applied to the estimates serve to determine the effect that chance can play but do not, in and of themselves, provide "proof" of causality.

We begin in Section 14.1 by discussing criteria by which to assess the quality of an estimate of the treatment effect. Some of this section repeats material first presented in Section 5.1; however, the emphasis now is on considerations applicable after the study has been completed and the first analysis has been done. In Section 14.2 we present six criteria which can help in assessing causality: (*a*) the *strength* of the association, (*b*) *consistency* of the association, (*c*) *specificity* of the association, (*d*) *temporality*, (*e*) *gradient* of effect, and (*f*) *coherence*. The discussion of these criteria is brief since they are meant to be used as a "checklist."

14.1 ASSESSING THE QUALITY OF THE ESTIMATE OF THE TREATMENT EFFECT

The major criterion for assessing the quality of the estimate of the treatment effect is the adequacy of the model. All the results presented concerning the expected amount of bias reduction are conditional on there being no model misspecification (i.e., on the assumptions such as linearity or parallelism being satisfied). These assumptions can often be tested. For example, the chi-square goodness-of-fit test can be used to test the fit of a log-linear model and the F test can be used to test for the parallelism of slopes for the ANCOVA model.

The lack of model misspecification also implies that all important confounding variables have been taken into account; that is, all the variables which have a significant effect on the estimate of the treatment effect. If a variable is suspected of confounding and data are available, the importance of the variable can be checked by including it in the chosen analysis. Its importance can be evaluated either in terms of the change in the estimate of the treatment effect or, in parametric analyses, by testing the statistical significance of the suspected confounder.

Another way to check for model misspecification in the case of nonmatching techniques is to compare the estimate of the treatment effect from the unadjusted data to that from the chosen analysis. A large difference in magnitude between these two estimates, while indicating that an adjustment procedure is needed, also indicates that the final estimate may be strongly influenced by the choice of adjustment procedure.

In addition to parametric and confirmatory statistical techniques, an "exploratory analysis" in the manner of Tukey (1977) is sometimes useful. His methods do not rely as heavily on parametric models and specific distributional assumptions.

14.2 ASSESSING CAUSALITY

Once the magnitude and direction of an association has been determined as accurately as possible, the previously listed six criteria can be considered before making any causal inferences. Each of these criteria is important in assessing the plausability of a cause-and-effect hypothesis, but the lack of any or all of them being satisfied does not reject the hypothesis.

The first criterion is the *strength* of the association or the magnitude of the treatment effect. The stronger the association, the stronger the evidence for causation. Consider the results from the studies of lung cancer and smoking. The death rate from lung cancer in cigarette smokers is 9 to 10 times the rate for nonsmokers, and in the case of heavy smokers the rate is 20 to 30 times as great. In the case of such a strong association, it is difficult to dismiss the likelihood of causality.

One can sometimes examine the strength of an association by assessing the effects of possible confounding variables. Under some special distributional assumptions, Bross (1966, and 1967) and Schlesselman (1978) have discussed methods to determine whether an apparent association is due in part or in whole to a variable that was not measured or controlled for in the study. (See Section 5.1 for a discussion of the "size rule.")

The second criterion is *consistency* of the association. Has the association been observed by several investigators at different times and places? As an example of a consistent association, observed under varied circumstances, consider the findings of the Advisory Committee's Report to the Surgeon-General of the United States (United States Department of Health Education and Welfare (1964), *Smoking and Health: Report of the Advisory Committee to the Surgeon-General of the Public Health Service,* Washington, DC: Public Health Service Publication No. 1103). The Committee found an association between lung cancer and smoking in 29 of the 29 case-control studies and in 7 of the 7 cohort studies.

An association that has been observed under a wide variety of circumstances lends strength to a cause-and-effect hypothesis. After all, the only way to determine whether a statistically significant result was indeed not due to only chance is by repetition. There are some situations, however, which cannot be repeated, and yet conclusions concerning cause and effect can be made. Hill (1965) discusses the experience of nickel refiners in South Wales with respect to lung cancer and nasal sinus cancer. The large number of cancer deaths in the population of workers in the nickel refineries was far above that expected in the general population. With such a large difference in death rates, it was clear, even without any biological information to support the statistical evidence, that working in these nickel refineries was a grave hazard to health.

The third criterion is *specificity* of association. Has the association been limited to only certain populations, areas, jobs, and so on? If an association appears only for specific segments of a population, say only in nickel refinery workers, and not in other segments, then causation is easier to infer. However, the lack of specificity does not negate the possibility of cause and effect.

Specificity can be sharpened by refining and subdividing the variables of interest. As an example, consider again the Advisory Committee's report on smoking in *Smoking and Health*. The overall analysis showed an increase in the risk of death from all causes in smokers. However, specificity was increased and causal inference improved when causes of death such as cancer were subdivided into specific-site cancers. By examining the association between smoking and forms of cancer, the specificity of smoking and lung cancer become apparent. In addition, smoking was refined by considering the method and frequency of smoking. This allowed the various researchers to identify cigarettes as a major causal factor.

The fourth criterion is *temporality* or time sequence of events. This criterion is most important in diseases or processes which take a long time to develop. In order to assign cause to any variable, it is necessary to know that the exposure to the potential causal variable happened sufficiently before the effect or response to indeed be considered causal. In some situations, it is a case of the chicken or the egg—which came first? For example, as Hill (1965) cites: "Does a particular diet lead to a disease or do the early stages of the disease lead to that particular dietetic routine?"

Both cohort and case-control types of studies suffer from problems in determining temporality. In a cohort study based on nonrandom assignment, one can sometimes argue that an apparent association may be due to the fact that, in the words of Susser (1973), "pre-existing attributes lead the cases to select themselves for the supposedly casual experience." Fisher (1958) proposed such an argument in the smoking and lung cancer controversery. He suggested that persons may be genetically predisposed to lung cancer as well as smoking. In case-control studies, however, since the information is obtained in retrospect, the precedence of the various factors may not be ascertainable. For example, unless a patient's medical records are available, the time of a patient's exposure to a suspected causal medication may not be ascertainable. The investigator would have to rely on a patient's potentially faulty recall.

The fifth criterion is the existence of a biological *gradient* or dose–response relationship. Causal inference can be strengthened if a monotonic relationship exists between the effect or response and the level of the hypothesized causal variable. In the smoking and lung cancer studies, the death rate from lung cancer rose linearly with the number of cigarettes smoked daily. Combining this information with the higher death rate for smokers than nonsmokers provided strong causal evidence. While the existence of an ascertainable dose–response relationship will strengthen causal inference, the lack of such a relationship will not weaken the hypothesis. A relationship may exist but in fact be so complex that it cannot be determined from the observed levels of the hypothesized causal variable.

The final criterion is *coherence* of the association. Do the results conflict with known facts concerning the development of a disease or condition being studied? If the results do conflict, however, one need not dismiss the apparent association, since the coherence of the result is conditional on the present state of knowledge.

For further more detailed discussion of these six criteria and additional criteria, the reader is referred to Hill (1965) and Susser (1973).

Our final advice to the investigator is to maintain some degree of skepticism for the duration of the study, so that objectively one can attempt to eliminate all other plausible explanations of a hypothesized result other than one's own. In the words of Sherlock Holmes, "when you have eliminated the impossible, whatever remains, *however improbable,* must be the truth" (Sir Arthur Conan Doyle, *The Sign of Four,* Chapt. 6).

REFERENCES

Bross, I. D. J. (1966), Spurious Effects from an Extraneous Variable, *Journal of Chronic Diseases,* **19,** 637–647.

Bross, I. D. J. (1967), Pertinency of an Extraneous Variable, *Journal of Chronic Diseases,* **20,** 487–495.

Fisher, R. A. (1958), Lung Cancer and Cigarettes? *Nature,* **182,** 108.

Hill, A. B. (1965), The Environment and Disease: Association or Causation? *Proceedings of the Royal Society of Medicine,* **58,** 295–300.

Schlesselman, J. J. (1978), Assessing Effects of Confounding Variables, *American Journal of Epidemiology,* **108,** 3–8.

Susser, M. (1973), *Causal Thinking in the Health Sciences, Concepts and Strategies of Epidemology,* New York: Oxford University Press.

Tukey, J. W. (1977), *Exploratory Data Analysis,* Reading, MA: Addison-Wesley.

Index